Graduate Texts in Mathematics 2

D0761582

John C. Oxtoby

Measure and Category

A Survey of the Analogies between
Topological and Measure Spaces

nd Edition

Mathematics)

form

eBook)

John C. Oxtoby
Department of Mathematics
Bryn Mawr College
Bryn Mawr, PA 19010
USA

AMS Subject Classification (1980):
26 A 21, 28 A 05, 54 C 50, 54 E 50, 54 H 05, 26-01, 28-01, 54-01

Library of Congress Cataloging in Publication Data

Oxtoby, John C.
  Measure and category.

  (Graduate texts in mathematics; 2)
  Bibliography: p.
  Includes index.
  1. Measure theory. 2. Topological spaces. 3. Categories (
I. Title. II. Series. QA312.09   1980   515.4′2   80-1577(

© 1971, 1980 by Springer-Verlag New York Inc.
Softcover reprint of the hardcover 2nd edition 1971

9 8 7 6 5 4 3 2 1

ISBN 978-1-4684-9341-2   ISBN 978-1-4684-9339-9
DOI 10.1007/978-1-4684-9339-9

Preface to the Second Edition

In this edition, a set of Supplementary Notes and Remarks has been added at the end, grouped according to chapter. Some of these call attention to subsequent developments, others add further explanation or additional remarks. Most of the remarks are accompanied by a briefly indicated proof, which is sometimes different from the one given in the reference cited. The list of references has been expanded to include many recent contributions, but it is still not intended to be exhaustive.

Bryn Mawr, April 1980   John C. Oxtoby

Preface to the First Edition

This book has two main themes: the Baire category theorem as a method for proving existence, and the "duality" between measure and category. The category method is illustrated by a variety of typical applications, and the analogy between measure and category is explored in all of its ramifications. To this end, the elements of metric topology are reviewed and the principal properties of Lebesgue measure are derived. It turns out that Lebesgue integration is not essential for present purposes—the Riemann integral is sufficient. Concepts of general measure theory and topology are introduced, but not just for the sake of generality. Needless to say, the term "category" refers always to Baire category; it has nothing to do with the term as it is used in homological algebra.

A knowledge of calculus is presupposed, and some familiarity with the algebra of sets. The questions discussed are ones that lend themselves naturally to set-theoretical formulation. The book is intended as an introduction to this kind of analysis. It could be used to supplement a standard course in real analysis, as the basis for a seminar, or for independent study. It is primarily expository, but a few refinements of known results are included, notably Theorem 15.6 and Proposition 20.4. The references are not intended to be complete. Frequently a secondary source is cited where additional references may be found.

The book is a revised and expanded version of notes originally prepared for a course of lectures given at Haverford College during the spring of 1957 under the auspices of the William Pyle Philips Fund. These, in turn, were based on the Earle Raymond Hedrick Lectures presented at the Summer Meeting of the Mathematical Association of America at Seattle, Washington, in August, 1956.

Bryn Mawr, April 1971   John C. Oxtoby

Contents

1. Measure and Category on the Line

The notions of measure and category are based on that of countability. Cantor's theorem, which says that no interval of real numbers is countable, provides a natural starting point for the study of both measure and category. Let us recall that a set is called *denumerable* if its elements can be put in one-to-one correspondence with the natural numbers $1, 2, \ldots$. A *countable* set is one that is either finite or denumerable. The set of rational numbers is denumerable, because for each positive integer k there are only a finite number ($\leq 2k - 1$) of rational numbers p/q in reduced form ($q > 0$, p and q relatively prime) for which $|p| + q = k$. By numbering those for which $k = 1$, then those for which $k = 2$, and so on, we obtain a sequence in which each rational number appears once and only once. Cantor's theorem reads as follows.

Theorem 1.1 (Cantor). *For any sequence $\{a_n\}$ of real numbers and for any interval I there exists a point p in I such that $p \neq a_n$ for every n.*

One proof runs as follows. Let I_1 be a closed subinterval of I such that $a_1 \notin I_1$. Let I_2 be a closed subinterval of I_1 such that $a_2 \notin I_2$. Proceeding inductively, let I_n be a closed subinterval of I_{n-1} such that $a_n \notin I_n$. The nested sequence of closed intervals I_n has a non-empty intersection. If $p \in \bigcap I_n$, then $p \in I$ and $p \neq a_n$ for every n.

This proof involves infinitely many unspecified choices. To avoid this objection the intervals must be chosen according to some definite rule. One such rule is this: divide I_{n-1} into three subintervals of equal length and take for I_n the first one of these that does not contain a_n. If we take I_0 to be the closed interval concentric with I and half as long, say, then all the choices are specified, and we have a well defined function of $(I, a_1, a_2, \ldots)$ whose value is a point of I different from all the a_n.

The fact that no interval is countable is an immediate corollary of Cantor's theorem.

With only a few changes, the above proof becomes a proof of the Baire category theorem for the line. Before we can formulate this theorem we need some definitions. A set A is *dense in the interval I* if A has a non-empty intersection with every subinterval of I; it is called *dense* if it is

1

dense in the line R. A set A is *nowhere dense* if it is dense in no interval, that is, if every interval has a subinterval contained in the complement of A. A nowhere dense set may be characterized as one that is "full of holes." The definition can be stated in two other useful ways: A is nowhere dense if and only if its complement A' contains a dense open set, and if and only if $\bar{A}$ (or A^-, the closure of A) has no interior points. The class of nowhere dense sets is closed under certain operations, namely

Theorem 1.2. *Any subset of a nowhere dense set is nowhere dense. The union of two (or any finite number) of nowhere dense sets is nowhere dense. The closure of a nowhere dense set is nowhere dense.*

Proof. The first statement is obvious. To prove the second, note that if A_1 and A_2 are nowhere dense, then for each interval I there is an interval $I_1 \subset I - A_1$ and an interval $I_2 \subset I_1 - A_2$. Hence $I_2 \subset I - (A_1 \cup A_2)$. This shows that $A_1 \cup A_2$ is nowhere dense. Finally, any open interval contained in A' is also contained in $A^{-\prime}$. ⬜

A denumerable union of nowhere dense sets is not in general nowhere dense, it may even be dense. For instance, the set of rational numbers is dense, but it is also a denumerable union of singletons (sets having just one element), and singletons are nowhere dense in R.

A set is said to be of *first category* if it can be represented as a countable union of nowhere dense sets. A subset of R that cannot be so represented is said to be of *second category*. These definitions were formulated in 1899 by R. Baire [18, p. 48], to whom the following theorem is due.

Theorem 1.3 (Baire). *The complement of any set of first category on the line is dense. No interval in R is of first category. The intersection of any sequence of dense open sets is dense.*

Proof. The three statements are essentially equivalent. To prove the first, let $A = \bigcup A_n$ be a representation of A as a countable union of nowhere dense sets. For any interval I, let I_1 be a closed subinterval of $I - A_1$. Let I_2 be a closed subinterval of $I_1 - A_2$, and so on. Then $\bigcap I_n$ is a non-empty subset of $I - A$, hence A' is dense. To specify all the choices in advance, it suffices to arrange the (denumerable) class of closed intervals with rational endpoints into a sequence, take $I_0 = I$, and for $n > 0$ take I_n to be the first term of the sequence that is contained in $I_{n-1} - A_n$.

The second statement is an immediate corollary of the first. The third statement follows from the first by complementation. ⬜

Evidently Baire's theorem implies Cantor's. Its proof is similar, although a different rule for choosing I_n was needed.

2

Theorem 1.4. *Any subset of a set of first category is of first category. The union of any countable family of first category sets is of first category.*

It is obvious that the class of first category sets has these closure properties. However, the closure of a set of first category is not in general of first category. In fact, the closure of a linear set A is of first category if and only if A is nowhere dense.

A class of sets that contains countable unions and arbitrary subsets of its members is called a σ-ideal. The class of sets of first category and the class of countable sets are two examples of σ-ideals of subsets of the line. Another example is the class of nullsets, which we shall now define.

The length of any interval I is denoted by $|I|$. A set $A \subset R$ is called a *nullset* (or a set of *measure zero*) if for each $\varepsilon > 0$ there exists a sequence of intervals I_n such that $A \subset \bigcup I_n$ and $\sum |I_n| < \varepsilon$.

It is obvious that singletons are nullsets and that any subset of a nullset is a nullset. Any countable union of nullsets is also a nullset. For suppose A_i is a nullset for $i = 1, 2, \ldots$. Then for each i there is a sequence of intervals I_{ij} ($j = 1, 2, \ldots$) such that $A_i \subset \bigcup_j I_{ij}$ and $\sum_j |I_{ij}| < \varepsilon/2^i$. The set of all the intervals I_{ij} covers A, and $\sum_{ij} |I_{ij}| < \varepsilon$, hence A is a nullset. This shows that the class of nullsets is a σ-ideal. Like the class of sets of first category, it includes all countable sets.

Theorem 1.5 (Borel). *If a finite or infinite sequence of intervals I_n covers an interval I, then $\sum |I_n| \geq |I|$.*

Proof. Assume first that $I = [a, b]$ is closed and that all of the intervals I_n are open. Let (a_1, b_1) be the first interval that contains a. If $b_1 \leq b$, let (a_2, b_2) be the first interval of the sequence that contains b_1. If $b_{n-1} \leq b$, let (a_n, b_n) be the first interval that contains b_{n-1}. This procedure must terminate with some $b_N > b$. Otherwise the increasing sequence $\{b_n\}$ would converge to a limit $x \leq b$, and x would belong to I_k for some k. All but a finite number of the intervals (a_n, b_n) would have to precede I_k in the given sequence, namely, all those for which $b_{n-1} \in I_k$. This is impossible, since no two of these intervals are equal. (Incidentally, this reasoning reproduces Borel's own proof of the "Heine-Borel theorem" [5, p. 228].) We have

$$b - a < b_N - a_1 = \sum_{i=2}^{N} (b_i - b_{i-1}) + b_1 - a_1 \leq \sum_{i=1}^{N} (b_i - a_i),$$

and so the theorem is true in this case.

In the general case, for any $\alpha > 1$ let J be a closed subinterval of I with $|J| = |I|/\alpha$, and let J_n be an open interval containing I_n with $|J_n| = \alpha|I_n|$. Then J is covered by the sequence $\{J_n\}$. We have already shown that

$\sum |J_n| \geqq |J|$. Hence $\alpha \sum |I_n| = \sum |J_n| \geqq |J| = |I|/\alpha$. Letting $\alpha \to 1$ we obtain the desired conclusion. ☐

This theorem implies that no interval is a nullset; it therefore provides still another proof of Cantor's theorem.

Every countable set is of first category and of measure zero. Some uncountable sets also belong to both classes. The simplest example is the *Cantor set* C, which consists of all numbers in the interval $[0, 1]$ that admit a ternary development in which the digit 1 does not appear. It can be constructed by deleting the open middle third of the interval $[0, 1]$, then deleting the open middle thirds of each of the intervals $[0, 1/3]$ and $[2/3, 1]$, and so on. If F_n denotes the union of the 2^n closed intervals of length $1/3^n$ which remain at the n-th stage, then $C = \bigcap F_n$. C is closed, since it is an intersection of closed sets. C is nowhere dense, since F_n (and therefore C) contains no interval of length greater than $1/3^n$. The sum of the lengths of the intervals that compose F_n is $(2/3)^n$, which is less than ε if n is taken sufficiently large. Hence C is a nullset. Finally, each number x in $(0, 1]$ has a unique non-terminating binary development $x = .x_1 x_2 x_3 \ldots$. If $y_i = 2x_i$, then $.y_1 y_2 y_3 \ldots$ is the ternary development with $y_i \neq 1$ of some point y of C. This correspondence between x and y, extended by mapping 0 onto itself, defines a one-to-one map of $[0, 1]$ onto a (proper) subset of C. It follows that C is uncountable; it has cardinality c (the power of the continuum).

The sets of measure zero and the sets of first category constitute two σ-ideals, each of which properly contains the class of countable sets. Their properties suggest that a set belonging to either class is "small" in one sense or another. A nowhere dense set is small in the intuitive geometric sense of being perforated with holes, and a set of first category can be "approximated" by such a set. A set of first category may or may not have any holes, but it always has a dense set of gaps. No interval can be represented as the union of a sequence of such sets. On the other hand, a nullset is small in the metric sense that it can be covered by a sequence of intervals of arbitrarily small total length. If a point is chosen at random in an interval in such a way that the probability of its belonging to any subinterval J is proportional to $|J|$, then the probability of its belonging to any given nullset is zero. It is natural to ask whether these notions of smallness are related. Does either class include the other? That neither class does, and that in some cases the two notions may be diametrically opposed, is shown by the following

Theorem 1.6. *The line can be decomposed into two complementary sets A and B such that A is of first category and B is of measure zero.*

Proof. Let $a_1, a_2, \ldots$ be an enumeration of the set of rational numbers (or of any countable dense subset of the line). Let I_{ij} be the open interval

4

with center a_i and length $1/2^{i+j}$. Let $G_j = \bigcup_{i=1}^{\infty} I_{ij}$ $(j = 1, 2, \ldots)$ and $B = \bigcap_{j=1}^{\infty} G_j$. For any $\varepsilon > 0$ we can choose j so that $1/2^j < \varepsilon$. Then $B \subset \bigcup_i I_{ij}$ and $\sum_i |I_{ij}| = \sum_i 1/2^{i+j} = 1/2^j < \varepsilon$. Hence B is a nullset. On the other hand, G_j is a dense open subset of R, since it is the union of a sequence of open intervals and it includes all rational points. Therefore its complement G'_j is nowhere dense, and $A = B' = \bigcup_j G'_j$ is of first category. $\square$

Corollary 1.7. *Every subset of the line can be represented as the union of a nullset and a set of first category.*

There is of course nothing paradoxical in the fact that a set that is small in one sense may be large in some other sense.

2. Liouville Numbers

Cantor's theorem, Baire's theorem, and Borel's theorem are existence theorems. If one can show that the set of numbers in an interval that lack a certain property is either countable, or a nullset, or a set of first category, then it follows that there exist points of the interval that have the property in question, in fact, most points of the interval (in the sense of cardinal number, or measure, or category, respectively) have the property. As a first illustration of this method let us consider the existence of transcendental numbers.

A complex number z is called *algebraic* if it satisfies some equation of the form

$$a_0 + a_1 z + a_2 z^2 + \cdots + a_n z^n = 0$$

with integer coefficients, not all zero. The *degree* of an algebraic number z is the smallest positive integer n such that z satisfies an equation of degree n. For instance, any rational number is algebraic of degree 1, $\sqrt{2}$ is algebraic of degree 2, and $\sqrt{2} + \sqrt{3}$ is algebraic of degree 4. Any real number that is not algebraic is called *transcendental*. Do there exist transcendental numbers? In view of Cantor's theorem, this question is answered by the following

Theorem 2.1. *The set of real algebraic numbers is denumerable.*

Proof. Let us define the weight of a polynomial $f(x) = \sum_0^n a_i x^i$ to be the number $n + \sum_0^n |a_i|$. There are only a finite number of polynomials having a given weight. Arrange these in some order, say lexicographically (first in order of n, then in order of a_0, and so on). Every non-constant polynomial has a weight at least equal to 2. Taking the polynomials of weight 2 in order, then those of weight 3, and so on, we obtain a sequence $f_1, f_2, f_3, \ldots$ in which every polynomial of degree 1 or more appears just once. Each polynomial has at most a finite number of real zeros. Number the zeros of f_1 in order, then those of f_2, and so on, passing over any that have already been numbered. In this way we obtain a definite enumeration of all real algebraic numbers. The sequence is infinite because it includes all rational numbers. $\square$

This gives perhaps the simplest proof of the existence of transcendental numbers. It should be noted that it is not an indirect proof; when all the choices are fixed in advance the construction used to prove Theorem 1.1 defines a specific transcendental number in $[0, 1]$. It may be laborious to compute even a few terms of its decimal development, but in principle the number can be computed to any desired accuracy.

An older and more informative proof of the existence of transcendental numbers is due to Liouville. His proof is based on the following

Lemma 2.2. *For any real algebraic number z of degree $n > 1$ there exists a positive integer M such that*

$$\left| z - \frac{p}{q} \right| > \frac{1}{Mq^n}$$

for all integers p and q, $q > 0$.

Proof. Let $f(x)$ be a polynomial of degree n with integer coefficients for which $f(z) = 0$. Let M be a positive integer such that $|f'(x)| \leqq M$ whenever $|z - x| \leqq 1$. Then, by the mean value theorem,

(1) $\qquad |f(x)| = |f(z) - f(x)| \leqq M|z - x| \quad$ whenever $\quad |z - x| \leqq 1$.

Now consider any two integers p and q, with $q > 0$. We wish to show that $|z - p/q| > 1/Mq^n$. This is evidently true in case $|z - p/q| > 1$, so we may assume that $|z - p/q| \leqq 1$. Then, by (1), $|f(p/q)| \leqq M|z - p/q|$, and therefore

(2) $\qquad\qquad |q^n f(p/q)| \leqq Mq^n|z - p/q|$.

The equation $f(x) = 0$ has no rational root (otherwise z would satisfy an equation of degree less then n). Moreover, $q^n f(p/q)$ is an integer. Hence the left member of (2) is at least 1 and we infer that $|z - p/q| \geqq 1/Mq^n$. Equality cannot hold, because z is irrational. $\quad\square$

A real number z is called a *Liouville number* if z is irrational and has the property that for each positive integer n there exist integers p and q such that

$$\left| z - \frac{p}{q} \right| < \frac{1}{q^n} \quad \text{and} \quad q > 1$$.

For example, $z = \sum_1^\infty 1/10^{k!}$ is a Liouville number. (Take $q = 10^{n!}$.)

Theorem 2.3. *Every Liouville number is transcendental.*

Proof. Suppose some Liouville number z is algebraic, of degree n. Then $n > 1$, since z is irrational. By Lemma 2.2 there exists a positive integer M such that

(3) $\qquad\qquad |z - p/q| > 1/Mq^n$

for all integers p and q with $q > 0$. Choose a positive integer k such that $2^k \geqq 2^n M$. Because z is a Liouville number there exist integers p and q,

with $q > 1$, such that
$$(4) \qquad |z - p/q| < 1/q^k.$$
From (3) and (4) it follows that $1/q^k > 1/Mq^n$. Hence $M > q^{k-n} \geq 2^{k-n} \geq M$, a contradiction. $\square$

Let us examine the set E of Liouville numbers. From the definition it follows at once that
$$(5) \qquad E = Q' \cap \bigcap_{n=1}^{\infty} G_n,$$
where Q denotes the set of rational numbers and
$$G_n = \bigcup_{q=2}^{\infty} \bigcup_{p=-\infty}^{\infty} (p/q - 1/q^n, p/q + 1/q^n).$$
G_n is a union of open intervals. Moreover, G_n includes every number of the form p/q, $q \geq 2$, hence $G_n \supset Q$. Therefore G_n is a dense open set, and so its complement is nowhere dense. Since, by (5), $E' = Q \cup \bigcup_{n=1}^{\infty} G'_n$, it follows that E' is of first category. Thus Baire's theorem implies that Liouville transcendental numbers exist in every interval, they are the "general case" in the sense of category.

What about the measure of E? From (5) it follows that $E \subset G_n$ for every n. Let
$$G_{n,q} = \bigcup_{p=-\infty}^{\infty} (p/q - 1/q^n, p/q + 1/q^n) \qquad (q = 2, 3, \ldots).$$
For any two positive integers m and n we have
$$E \cap (-m, m) \subset G_n \cap (-m, m)$$
$$= \bigcup_{q=2}^{\infty} [G_{n,q} \cap (-m, m)] \subset \bigcup_{q=2}^{\infty} \bigcup_{p=-mq}^{mq} (p/q - 1/q^n, p/q + 1/q^n).$$
Therefore $E \cap (-m, m)$ can be covered by a sequence of intervals the sum of whose lengths, for any $n > 2$, is
$$\sum_{q=2}^{\infty} \sum_{p=-mq}^{mq} 2/q^n = \sum_{q=2}^{\infty} (2mq + 1)(2/q^n) \leq \sum_{q=2}^{\infty} (4mq + q)(1/q^n)$$
$$= (4m+1) \sum_{q=2}^{\infty} 1/q^{n-1} \leq (4m+1) \int_1^{\infty} \frac{dx}{x^{n-1}} = \frac{4m+1}{n-2}.$$
It follows that $E \cap (-m, m)$ is a nullset for every m, and therefore E is a nullset.

Thus E is small in the sense of measure, but large in the sense of category. The sets E and E' provide another decomposition of the line into a set of measure zero and a set of first category (cf. Theorem 1.6). Moreover, the set E is small in an even stronger sense, as we shall now show.

If s is a positive real number and $E \subset R$, then E is said to have *s-dimensional Hausdorff measure zero* if for each $\varepsilon > 0$ there is a sequence of intervals I_n such that $E \subset \bigcup_{n=1}^{\infty} I_n$, $\sum_{n=1}^{\infty} |I_n|^s < \varepsilon$, and $|I_n| < \varepsilon$ for every n. The sets of s-dimensional measure zero constitute a σ-ideal. For $s = 1$ it coincides with the class of nullsets, and for $0 < s < 1$ it is a proper subclass. The following theorem therefore strengthens the proposition that E is a nullset.

Theorem 2.4. *The set E of Liouville numbers has s-dimensional Hausdorff measure zero, for every $s > 0$.*

Proof. It suffices to find, for each $\varepsilon > 0$ and for each positive integer m, a sequence of intervals I_n such that

$$E \cap (-m, m) \subset \bigcup_{n=1}^{\infty} I_n, \quad \sum_{n=1}^{\infty} |I_n|^s < \varepsilon, \quad \text{and} \quad |I_n| < \varepsilon.$$

For each positive integer n, we have

$$E \cap (-m, m) \subset \bigcup_{q=2}^{\infty} \bigcup_{p=-mq}^{mq} (p/q - 1/q^n, p/q + 1/q^n).$$

Choose n so as to satisfy simultaneously the following conditions:

$$1/2^{n-1} < \varepsilon, \quad ns > 2, \quad \frac{(2m+1) \, 2^s}{ns - 2} < \varepsilon.$$

Then each of the intervals $(p/q - 1/q^n, p/q + 1/q^n)$ has length $2/q^n \leqq 2/2^n < \varepsilon$, and we have

$$\sum_{q=2}^{\infty} \sum_{p=-mq}^{mq} (2/q^n)^s = \sum_{q=2}^{\infty} \frac{(2mq + 1) \, 2^s}{q^{ns}}$$

$$\leqq (2m+1) \, 2^s \sum_{q=2}^{\infty} \frac{1}{q^{ns-1}} \leqq (2m+1) \, 2^s \int_1^{\infty} \frac{dx}{x^{ns-1}}$$

$$= \frac{(2m+1) \, 2^s}{ns - 2} < \varepsilon. \quad \square$$

3. Lebesgue Measure in r-Space

By an *interval* I in Euclidean r-space $(r=1, 2,...)$ is meant a rectangular parallelepiped with edges parallel to the axes. It is the Cartesian product of r 1-dimensional intervals. As in the 1-dimensional case, the r-dimensional volume of I will be denoted by $|I|$. Lebesgue measure in r-space is an extension of the notion of volume to a larger class of sets. Thus Lebesgue measure has a different meaning in spaces of different dimension. However, since we shall usually regard the dimension as fixed, there is no need to indicate r explicitly in our notations.

A sequence of intervals I_i is said to *cover* the set A if its union contains A. The greatest lower bound of the sums $\sum|I_i|$, for all sequences $\{I_i\}$ that cover A, is called the *outer measure* of A; it is denoted by $m^*(A)$. Thus for any subset A of r-space,

$$m^*(A) = \inf\{\sum|I_i| : A \subset \bigcup I_i\}\ .$$

When A belongs to a certain class of sets to be defined presently, $m^*(A)$ will be called the Lebesgue measure of A, and denoted by $m(A)$.

The edges of the intervals I_i may be closed, open, or half-open, and the sequence of intervals may be finite or infinite. It may happen that the series $\sum|I_i|$ diverges for every sequence $\{I_i\}$ that covers A; in this case $m^*(A) = \infty$. In all other cases $m^*(A)$ is a nonnegative real number.

This definition can be modified in either or both of two respects without affecting the value of $m^*(A)$. In the first place, we may require that the diameters of the intervals I_i should all be less than a given positive number δ. This is clear since each interval can be divided into subintervals of diameter less than δ without affecting the sum of their volumes. Secondly, we may require that all the intervals be open. For any covering sequence $\{I_i\}$ and $\varepsilon > 0$ we can find open intervals J_i such that $I_i \subset J_i$ and $\sum|J_i| \leq \sum|I_i| + \varepsilon$. Hence the greatest lower bound for open coverings is the same as for all covering sequences.

We shall now deduce a number of properties of outer measure.

Theorem 3.1. *If* $A \subset B$ *then* $m^*(A) \leq m^*(B)$.

This is obvious, since any sequence $\{I_i\}$ that covers B also covers A.

10

Theorem 3.2. *If* $A = \bigcup A_i$ *then* $m^*(A) \leq \sum m^*(A_i)$.

This property of outer measure is called *countable subadditivity*. For any $\varepsilon > 0$ there is a sequence of intervals $I_{i,j}$ ($j = 1, 2, \ldots$) that covers A_i such that $\sum_j |I_{i,j}| \leq m^*(A_i) + \varepsilon/2^i$. Then $A \subset \bigcup_{i,j} I_{i,j}$ and $\sum_{i,j} |I_{i,j}| \leq \sum_i m^*(A_i) + \varepsilon$. Therefore $m^*(A) \leq \sum m^*(A_i) + \varepsilon$. Letting $\varepsilon \to 0$, the required inequality follows.

Theorem 3.3. *For any interval* I, $m^*(I) = |I|$.

Proof. The inequality $m^*(I) \leq |I|$ is clear, since I covers itself. To prove the inverse inequality, let ε be an arbitrary positive number and let $\{I_i\}$ be an open covering of I such that $\sum |I_i| < m^*(I) + \varepsilon$. Let J be a closed subinterval of I such that $|J| > |I| - \varepsilon$. By the Heine-Borel theorem, $J \subset \bigcup_1^k I_i$ for some k. Let $K_1, \ldots, K_n$ be an enumeration of the closed intervals into which $\bar{I}_1, \ldots, \bar{I}_k$ are divided by all the $(r-1)$-dimensional hyperplanes that contain an $(r-1)$-dimensional face of one of the intervals $I_1, \ldots, I_k$, or J, and let $J_1, \ldots, J_m$ be the closed intervals into which J is divided by these same hyperplanes. Then each interval J_i is equal to at least one of the intervals K_j. Consequently,

$$|J| = \sum_{i=1}^m |J_i| \leq \sum_{j=1}^n |K_j| = \sum_{i=1}^k |I_i| < m^*(I) + \varepsilon.$$

Therefore $|I| \leq m^*(I) + 2\varepsilon$. The desired inequality follows by letting $\varepsilon \to 0$. []

Generalizing the definition given in Chapter 1, any subset of r-space with outer measure zero is called a *nullset*, or *set of measure zero*. A statement that holds for all points of a set E except a set of measure zero is said to hold *almost everywhere*, or for *almost all* points of E.

We next deduce some results which are included in later theorems. Accordingly, we designate them as lemmas.

Lemma 3.4. *If* F_1 *and* F_2 *are disjoint bounded closed sets, then* $m^*(F_1 \cup F_2) = m^*(F_1) + m^*(F_2)$.

Proof. There is a positive number δ such that no interval of diameter less than δ meets both F_1 and F_2. For any $\varepsilon > 0$ there is a sequence of intervals I_i of diameter less than δ such that $F_1 \cup F_2 \subset \bigcup I_i$ and $\sum |I_i| \leq m^*(F_1 \cup F_2) + \varepsilon$. Let $\sum' |I_i|$ denote the sum over those intervals that meet F_1, and let $\sum'' |I_i|$ denote the sum over the remaining intervals (which cover F_2). Then

$$m^*(F_1) + m^*(F_2) \leq \sum' |I_i| + \sum'' |I_i| = \sum |I_i| \leq m^*(F_1 \cup F_2) + \varepsilon.$$

Letting $\varepsilon \to 0$, we conclude that

$$m^*(F_1) + m^*(F_2) \leq m^*(F_1 \cup F_2).$$

The reverse inequality follows from Theorem 3.2. []

Lemma 3.5. *If $F_1, ..., F_n$ are disjoint bounded closed sets, then* $m^*(\bigcup_1^n F_i) = \sum_1^n m^*(F_i)$.

This follows from Lemma 3.4 by induction on n.

Lemma 3.6. *For any bounded open set G and $\varepsilon > 0$ there exists a closed set F such that $F \subset G$ and $m^*(F) > m^*(G) - \varepsilon$.*

Proof. G can be represented as the union of a sequence of non-overlapping intervals I_i. By definition, $m^*(G) \leq \sum |I_i|$. Determine n so that $\sum_1^n |I_i| > m^*(G) - \varepsilon/2$, and let J_i be a closed interval contained in the interior of I_i such that $|J_i| > |I_i| - \varepsilon/2n$ $(i = 1, 2, ..., n)$. Then $F = \bigcup_1^n J_i$ is a closed subset of G, and by Theorem 3.3 and Lemma 3.5, $m^*(F) = \sum_1^n |J_i| > \sum_1^n |I_i| - \varepsilon/2 > m^*(G) - \varepsilon$. □

Lemma 3.7. *If F is a closed subset of a bounded open set G, then* $m^*(G - F) = m^*(G) - m^*(F)$.

Proof. By Lemma 3.6, for any $\varepsilon > 0$ there is a closed subset F_1 of the open set $G - F$ such that $m^*(F_1) > m^*(G - F) - \varepsilon$. By Lemma 3.4 and Theorem 3.1,

$$m^*(F) + m^*(G - F) < m^*(F) + m^*(F_1) + \varepsilon = m^*(F \cup F_1) + \varepsilon \leq m^*(G) + \varepsilon.$$

Letting $\varepsilon \to 0$, we conclude that

$$m^*(F) + m^*(G - F) \leq m^*(G).$$

The reverse inequality follows from Theorem 3.2. □

Definition 3.8. A set A is *measurable* (in the sense of Lebesgue) if for each $\varepsilon > 0$ there exists a closed set F and an open set G such that $F \subset A \subset G$ and $m^*(G - F) < \varepsilon$.

Lemma 3.9. *If A is measurable, then A' is measurable.*

For if $F \subset A \subset G$, then $F' \supset A' \supset G'$ and $F' - G' = G - F$.

Lemma 3.10. *If A and B are measurable, then $A \cap B$ is measurable.*

Proof. Let F_1 and F_2 be closed sets, and let G_1 and G_2 be open sets, such that $F_1 \subset A \subset G_1$, $F_2 \subset B \subset G_2$, $m^*(G_1 - F_1) < \varepsilon/2$, $m^*(G_2 - F_2) < \varepsilon/2$. Then $F = F_1 \cap F_2 \subset A \cap B \subset G_1 \cap G_2 = G$, say, and

$$G - F \subset (G_1 - F_1) \cup (G_2 - F_2).$$

Hence $m^*(G - F) \leq m^*(G_1 - F_1) + m^*(G_2 - F_2) < \varepsilon$. □

Lemma 3.11. *A bounded set A is measurable if for each $\varepsilon > 0$ there exists a closed set $F \subset A$ such that $m^*(F) > m^*(A) - \varepsilon$.*

Proof. For any $\varepsilon > 0$ let F be a closed subset of A such that $m^*(F) > m^*(A) - \varepsilon/2$. Since $m^*(A) < \infty$ there exists a covering sequence of open intervals I_i of diameter less than 1 such that $\sum |I_i| < m^*(A) + \varepsilon/2$. Let G be the union of those intervals I_i that meet A. Then $F \subset A \subset G$, G is bounded, and by Lemma 3.7, $m^*(G - F) = m^*(G) - m^*(F) \leq \sum |I_i| - m^*(F) < m^*(A) + \varepsilon/2 - m^*(F) < \varepsilon$. Hence A is measurable. $\quad\Box$

Lemma 3.12. *Any interval and any nullset is measurable.*

Proof. The first statement follows at once from Lemma 3.11 and Theorem 3.3. If $m^*(A) = 0$, then for each $\varepsilon > 0$ there is a covering sequence of open intervals I_i such that $\sum |I_i| < \varepsilon$. Take $G = \bigcup I_i$ and $F = \emptyset$. Then F is closed, G is open, $F \subset A \subset G$, and $m^*(G - F) \leq \sum |I_i| < \varepsilon$. Hence A is measurable. $\quad\Box$

Lemma 3.13. *Let $\{A_i\}$ be a disjoint sequence of measurable sets all contained in some interval I. If $A = \bigcup A_i$, then A is measurable and $m^*(A) = \sum m^*(A_i)$.*

Proof. For any $\varepsilon > 0$ there exist closed sets $F_i \subset A_i$ such that $m^*(F_i) > m^*(A_i) - \varepsilon/2^{i+1}$. By countable subadditivity, $m^*(A) \leq \sum_1^\infty m^*(A_i)$. Determine k so that

$$\sum_1^k m^*(A_i) > m^*(A) - \varepsilon/2 ,$$

and put $F = \bigcup_1^k F_i$. Then, by Lemma 3.5,

$$m^*(F) = \sum_1^k m^*(F_i) > \sum_1^k m^*(A_i) - \varepsilon/2 > m^*(A) - \varepsilon .$$

Hence A is measurable, by Lemma 3.11. For any n we have

$$\sum_1^n m^*(A_i) < \sum_1^n m^*(F_i) + \varepsilon/2 = m^*\left(\bigcup_1^n F_i\right) + \varepsilon/2 \leq m^*(A) + \varepsilon/2 .$$

Letting $n \to \infty$ and then letting $\varepsilon \to 0$ we conclude that $\sum_1^\infty m^*(A_i) \leq m^*(A)$. The reverse inequality follows from countable subadditivity. $\quad\Box$

Lemma 3.14. *For any disjoint sequence of measurable sets A_i, the set $A = \bigcup A_i$ is measurable and $m^*(A) = \sum m^*(A_i)$.*

Proof. Let I_j ($j = 1, 2, \ldots$) be a sequence of disjoint intervals whose union is the whole of r-space such that any bounded set is covered by finitely many. By Lemmas 3.10 and 3.12, the sets $A_{ij} = A_i \cap I_j$ are measurable. They are also disjoint. Put $B_j = \bigcup_i A_{ij}$. By Lemma 3.13, B_j is a measurable subset of I_j. The sets B_j are disjoint, and $A = \bigcup B_j$. For any $\varepsilon > 0$ there exist closed sets F_j and bounded open sets G_j such that $F_j \subset B_j \subset G_j$ and $m^*(G_j - F_j) < \varepsilon/2^j$. Let $F = \bigcup F_j$ and $G = \bigcup G_j$. Then F is closed, since any convergent sequence contained in F is bounded and therefore contained in the union of a finite number of the sets F_j, which is a closed subset of F. Also, G is open. We have $F \subset A \subset G$ and $G - F = \bigcup (G_j - F)$

$\subset \bigcup(G_j - F_j)$. Hence $m^*(G - F) \leq \sum m^*(G_j - F_j) < \varepsilon$. This shows that A is measurable. Since $A_i = \bigcup_j A_{ij}$, we have $m^*(A_i) \leq \sum_j m^*(A_{ij})$, and therefore

$$\sum_i m^*(A_i) \leq \sum_{i,j} m^*(A_{ij}) = \sum_j \sum_i m^*(A_{ij}) = \sum_j m^*(B_j),$$

by Lemma 3.13. Also, for any n,

$$\sum_1^n m^*(B_j) \leq \sum_1^n m^*(F_j) + \sum_1^n m^*(G_j - F_j)$$
$$\leq m^*(\bigcup_1^n F_j) + \varepsilon \leq m^*(A) + \varepsilon.$$

Letting $n \to \infty$ and then letting $\varepsilon \to 0$ we conclude that $\sum_j m^*(B_j) \leq m^*(A)$. Therefore $\sum m^*(A_i) \leq m^*(A)$. The reverse inequality again follows from countable subadditivity. □

We have now established the most important properties of outer measure. To formulate them conveniently, we need some additional definitions.

A non-empty class S of subsets of a set X is called a *ring* of subsets of X if it contains the union and the difference of any two of its members. It is called a *σ-ring* if it also contains the union of any sequence of its members. A ring (or σ-ring) of subsets of X is called an *algebra* (respectively, a *σ-algebra*) of subsets of X if X itself is a member of the ring. Evidently a class of subsets of X is an algebra if and only if it is closed under the operations of union (or intersection) and complementation; it is a σ-algebra if it is also closed under countable union (or countable intersection).

A set function μ defined on a ring S of subsets of X is said to be *countably additive* if the equation $\mu(A) = \sum \mu(A_i)$ holds whenever $\{A_i\}$ is a disjoint sequence of members of S whose union A also belongs to S. A *measure* is an extended real valued, non-negative, countably additive set function μ, defined on a σ-ring S of subsets of a set X, and such that $\mu(\emptyset) = 0$. A triple (X, S, μ), where S is a σ-ring of subsets of X and μ is a measure defined on S, is called a *measure space*. Sets belonging to S are called *μ-measurable*. When every subset of a set of μ-measure zero belongs to S (that is, when the sets of μ-measure zero constitute a σ-ideal), the measure space is said to be *complete*.

By Lemmas 3.9, 3.10, 3.12, and 3.14, the class S of measurable sets is a σ-algebra of subsets of r-space, and m^* is countably additive on S. Hence the restriction of m^* to S is a measure; it is called (r-dimensional) *Lebesgue measure* and is denoted by m. Since S includes all intervals, it follows that S includes all open sets, all closed sets, all F_σ sets (countable unions of closed sets), and all G_δ sets (countable intersections of open sets). Moreover,

Theorem 3.15. *A set A is measurable if and only if it can be represented as an F_σ set plus a nullset (or as a G_δ set minus a nullset).*

Proof. If A is measurable, then for each n there exists a closed set F_n and an open set G_n such that $F_n \subset A \subset G_n$ and $m^*(G_n - F_n) < 1/n$. Put $E = \bigcup F_n$ and $N = A - E$. Then E is an F_σ set. N is a nullset, since $N \subset G_n - F_n$ and $m^*(N) < 1/n$ for every n. A is the disjoint union of E and N. It follows by complementation that A can also be represented as a G_δ set minus a nullset. Conversely, any set that can be so represented is measurable, by Lemma 3.12 and the fact that S is a σ-algebra. $\square$

For any class of subsets of a set X there is a smallest σ-algebra of subsets of X that contains it, namely, the intersection of all such σ-algebras. This is called the *σ-algebra generated by the class.* The members of the σ-algebra of subsets of r-space generated by the class of open sets (or closed sets, or intervals) are called *Borel sets.* Hence every Borel subset of r-space is measurable. By Theorem 3.15, the Borel sets together with the nullsets generate the class of measurable sets. Summarizing, we have

Theorem 3.16. *The class S of measurable sets is the σ-algebra of subsets of r-space X generated by the open sets together with the nullsets. Lebesgue measure m is a measure on S such that $m(I) = |I|$ for every interval I. (X, S, m) is a complete measure space.*

The following theorem expresses the property of countable additivity in a form that is often more convenient.

Theorem 3.17. *If A_i is measurable, and $A_i \subset A_{i+1}$ for each i, then the set $A = \bigcup A_i$ is measurable and $m(A) = \lim m(A_i)$. If A_i is measurable and $A_i \supset A_{i+1}$ for each i, then the set $A = \bigcap A_i$ is measurable, and $m(A) = \lim m(A_i)$ provided $m(A_i) < \infty$ for some i.*

Proof. In the first case, put $B_1 = A_1$ and $B_i = A_i - A_{i-1}$ for $i > 1$. Then $\{B_i\}$ is a disjoint sequence of measurable sets, with $A = \bigcup B_i$. Hence

$$m(A) = \sum m(B_i) = \lim \sum_1^n m(B_i) = \lim m(A_n),$$

where the limit may be equal to ∞.

In the second case, we may assume $m(A_1) < \infty$. Put $B_i = A_1 - A_i$ and $B = A_1 - A$. Then $B_i \subset B_{i+1}$ and $\bigcup B_i = B$. Hence $m(A_1) - m(A) = m(B) = \lim m(B_i) = \lim (m(A_1) - m(A_i)) = m(A_1) - \lim m(A_i)$, and so $m(A) = \lim m(A_i)$, both members being finite. $\square$

The manner in which the set function m^* is determined by its values on closed and open sets is indicated by the following

Theorem 3.18. *The outer measure of any set A is expressed by the formula*

$$m^*(A) = \inf \{m(G) : A \subset G, \ G \ \text{open}\}.$$

15

If A is measurable, then

$$m^*(A) = \sup\{m(F) : A \supset F, \ F \text{ bounded and closed}\}.$$

Conversely, if this equation holds and $m^*(A) < \infty$, then A is measurable.

Proof. The first statement is clear, since the union of any covering sequence of open intervals is an open superset of A. To prove the second, let α be any real number less than $m(A)$, and let $A_i = A \cap (-i, i)^r$. By Theorem 3.17, $m(A) = \lim m(A_i)$, hence we can choose i so that $m(A_i) > \alpha$. By measurability, A_i (which is bounded) contains a closed set F with $m(F) > \alpha$, and F is also a subset of A. Conversely, if $m^*(A) < \infty$ and F is a closed subset of A with $m(F) > m^*(A) - \varepsilon/2$, let G be an open superset of A such that $m(G) < m^*(A) + \varepsilon/2$. Then $F \subset A \subset G$ and $m(G - F) < \varepsilon$, hence A is measurable. $\square$

It may be noted that Lemmas 3.4 through 3.14 are implicitly included in Theorems 3.16 and 3.18.

The following theorem expresses the fact that Lebesgue measure is invariant under translation.

Theorem 3.19. *If A is congruent by translation to a measurable set B, then A is measurable and $m(A) = m(B)$.*

This is clear from the definitions and from the fact that congruent intervals have equal volume. Measurability and measure are also preserved by rotations and reflections of r-space, but we shall not prove this.

The definition of measurability, and the fact that any open set is the union of a sequence of disjoint intervals, implies that any set of finite measure can be obtained from some finite union of disjoint intervals by adding and subtracting two sets of arbitrarily small measure. So to speak, a set of finite measure is approximately equal to a finite union of intervals. Much deeper is the fact that a measurable set has locally a kind of all or none structure; at almost all points it is either highly concentrated or highly rarified. This idea is made precise by a remarkable theorem, due to Lebesgue, with which we conclude this chapter. We shall consider only the 1-dimensional case.

A measurable set $E \subset R$ is said to have *density d at x* if the

$$\lim_{h \to 0} \frac{m(E \cap [x - h, x + h])}{2h}$$

exists and is equal to d. Let us denote the set of points of R at which E has density 1 by $\phi(E)$. Then E has density 0 at each point of $\phi(R - E)$. ϕ is called the Lebesgue *lower density*. Lebesgue's theorem asserts that $\phi(E)$ is measurable and differs from E by a nullset. This implies that E has density 1 at almost all points of E, and density 0 at almost all points of $R - E$. Thus it is impossible, for instance, for a set and its complement each to include exactly half of the (outer) measure of every interval. (Such a set would be measurable and would have density 1/2 everywhere.)

The *symmetric difference* of two sets A and B is the set of points that belong to one but not to both of the sets. It is denoted by $A \triangle B$. Thus $A \triangle B = (A - B) \cup (B - A)$.

Theorem 3.20 (Lebesgue Density Theorem). *For any measurable set $E \subset R$, $m(E \triangle \phi(E)) = 0$.*

Proof. It is sufficient to show that $E - \phi(E)$ is a nullset, since $\phi(E) - E \subset E' - \phi(E')$ and E' is measurable. We may also assume that E is bounded. Furthermore, $E - \phi(E) = \bigcup_{\varepsilon > 0} A_\varepsilon$, where

$$A_\varepsilon = \left\{ x \in E : \liminf_{h \to 0} \frac{m(E \cap [x - h, x + h])}{2h} < 1 - \varepsilon \right\}.$$

Hence it is sufficient to show that A_ε is a nullset for every $\varepsilon > 0$. Putting $A = A_\varepsilon$, we shall obtain a contradiction from the supposition that $m^*(A) > 0$.

If $m^*(A) > 0$ there exists a bounded open set G containing A such that $m(G) < m^*(A)/(1 - \varepsilon)$. Let $\mathscr{E}$ denote the class of all closed intervals I such that $I \subset G$ and $m(E \cap I) \leq (1 - \varepsilon) |I|$. Observe that (i) $\mathscr{E}$ includes arbitrarily short intervals about each point of A, and (ii) for any disjoint sequence $\{I_n\}$ of members of $\mathscr{E}$, we have $m^*(A - \bigcup I_n) > 0$. Property (ii) follows from the fact that

$$m^*(A \cap \bigcup I_n) \leq \sum m(E \cap I_n) \leq (1 - \varepsilon) \sum |I_n| \leq (1 - \varepsilon) m(G) < m^*(A).$$

We construct inductively a disjoint sequence I_n of members of $\mathscr{E}$ as follows. Choose I_1 arbitrarily from $\mathscr{E}$. Having chosen $I_1, ..., I_n$, let $\mathscr{E}_n$ be the set of members of $\mathscr{E}$ that are disjoint to $I_1, ..., I_n$. Properties (i) and (ii) imply that $\mathscr{E}_n$ is non-empty. Let d_n be the least upper bound of the lengths of members of $\mathscr{E}_n$, and choose $I_{n+1} \in \mathscr{E}_n$ such that $|I_{n+1}| > d_n/2$. Put $B = A - \bigcup_1^\infty I_n$. By (ii), we have $m^*(B) > 0$. Hence there exists a positive integer N such that

$$(1) \qquad\qquad \sum_{N+1}^\infty |I_n| < m^*(B)/3 .$$

For each $n > N$ let J_n denote the interval concentric with I_n with $|J_n| = 3|I_n|$. The inequality (1) implies that the intervals $J_n (n > N)$ do not cover B, hence there exists a point $x \in B - \bigcup_{N+1}^\infty J_n$. Since $x \in A - \bigcup_1^N I_n$, it follows from (i) that there exists an interval $I \in \mathscr{E}_N$ with center x. I must meet some interval I_n with $n > N$. (Otherwise $|I| \leq d_n < 2|I_{n+1}|$ for all n, contrary to $\sum_1^\infty |I_n| \leq m(G) < \infty$.) Let k be the least integer such that I meets I_k. Then $k > N$ and $|I| \leq d_{k-1} < 2|I_k|$. It follows that the center x of I belongs to J_k, contrary to $x \notin \bigcup_{N+1}^\infty J_n$. $\quad\square$

17

Let us write $A \sim B$ when $m(A \triangle B) = 0$. This is an equivalence relation in the class S of measurable sets. The following theorem says that the mapping $\phi : S \to S$ may be regarded as a function that selects one member from each equivalence class. Moreover, it does so in such a way that the selected sets constitute a class that includes the empty set, the whole space, and is closed under intersection.

Theorem 3.21. *For any measurable set A, let $\phi(A)$ denote the set of points of R where A has density 1. Then ϕ has the following properties, where $A \sim B$ means that $A \triangle B$ is a nullset:*
1) $\phi(A) \sim A$,
2) $A \sim B$ *implies* $\phi(A) = \phi(B)$,
3) $\phi(\emptyset) = \emptyset$ *and* $\phi(R) = R$,
4) $\phi(A \cap B) = \phi(A) \cap \phi(B)$,
5) $A \subset B$ *implies* $\phi(A) \subset \phi(B)$.

Proof. The first assertion is just the Lebesgue density theorem. The second and third are immediate consequences of the definition of ϕ. To prove 4), note that for any interval I we have $I - (A \cap B) = (I - A) \cup (I - B)$. Hence $m(I) - m(I \cap A \cap B) \leq m(I) - m(I \cap A) + m(I) - m(I \cap B)$. Therefore

$$\frac{m(I \cap A)}{|I|} + \frac{m(I \cap B)}{|I|} - 1 \leq \frac{m(I \cap A \cap B)}{|I|}.$$

Taking $I = [x - h, x + h]$ and letting $h \to 0$ it follows that $\phi(A) \cap \phi(B) \subset \phi(A \cap B)$. The opposite inclusion is obvious. Property 5) is a consequence of 4). $\square$

4. The Property of Baire

The operation of symmetric difference, defined by

$$A \triangle B = (A \cup B) - (A \cap B) = (A - B) \cup (B - A),$$

is commutative, associative, and satisfies the distributive law $A \cap (B \triangle C)$ $= (A \cap B) \triangle (A \cap C)$. Evidently, $A \triangle B \subset A \cup B$ and $A \triangle A = \emptyset$. It is easy to verify that any class of sets that is closed under $\triangle$ and $\cap$ is a commutative ring (in the algebraic sense) when these operations are taken to define addition and multiplication, respectively. Such a class is also closed under the operations of union and difference. It is therefore a ring of subsets of its union, as this term was defined in Chapter 3.

A subset A of r-space (or of any topological space) is said to have the *property of Baire* if it can be represented in the form $A = G \triangle P$, where G is open and P is of first category.

Theorem 4.1. *A set A has the property of Baire if and only if it can be represented in the form $A = F \triangle Q$, where F is closed and Q is of first category.*

Proof. If $A = G \triangle P$, G open and P of first category, then $N = \bar{G} - G$ is a nowhere dense closed set, and $Q = N \triangle P$ is of first category. Let $F = \bar{G}$. Then $A = G \triangle P = (\bar{G} \triangle N) \triangle P = \bar{G} \triangle (N \triangle P) = F \triangle Q$. Conversely, if $A = F \triangle Q$, where F is closed and Q is of first category, let G be the interior of F. Then $N = F - G$ is nowhere dense, $P = N \triangle Q$ is of first category, and $A = F \triangle Q = (G \triangle N) \triangle Q = G \triangle (N \triangle Q) = G \triangle P.$ ☐

Theorem 4.2. *If A has the property of Baire, then so does its complement.*

Proof. For any two sets A and B we have $(A \triangle B)' = A' \triangle B$. Hence if $A = G \triangle P$, then $A' = G' \triangle P$, and the conclusion follows from Theorem 4.1. ☐

Theorem 4.3. *The class of sets having the property of Baire is a σ-algebra. It is the σ-algebra generated by the open sets together with the sets of first category.*

Proof. Let $A_i = G_i \triangle P_i$ $(i = 1, 2, \ldots)$ be any sequence of sets having the property of Baire. Put $G = \bigcup G_i$, $P = \bigcup P_i$, and $A = \bigcup A_i$. Then G is open, P is of first category, and $G - P \subset A \subset G \cup P$. Hence $G \triangle A \subset P$ is of first category, and $A = G \triangle (G \triangle A)$ has the property of Baire. This result, together with Theorem 4.2, shows that the class in question is a σ-algebra. It is evidently the smallest σ-algebra that includes all open sets and all sets of first category. $\square$

Theorem 4.4. *A set has the property of Baire if and only if it can be represented as a G_δ set plus a set of first category (or as an F_σ set minus a set of first category).*

Proof. Since the closure of any nowhere dense set is nowhere dense, any set of first category is contained in an F_σ set of first category. If G is open and P is of first category, let Q be an F_σ set of first category that contains P. Then the set $E = G - Q$ is a G_δ, and we have

$$G \triangle P = [(G - Q) \triangle (G \cap Q)] \triangle (P \cap Q)$$
$$= E \triangle [(G \triangle P) \cap Q].$$

The set $(G \triangle P) \cap Q$ is of first category and disjoint to E. Hence any set having the property of Baire can be represented as the disjoint union of a G_δ set and a set of first category. Conversely, any set that can be so represented belongs to the σ-algebra generated by the open sets and the sets of first category; it therefore has the property of Baire. The parenthetical statement follows by complementation, with the aid of Theorem 4.2. $\square$

A *regular open set* is a set that is equal to the interior of its closure. Any set of the form $A^{-\prime-\prime}$ is regular open.

Theorem 4.5. *Any open set H is of the form $H = G - \bar{N}$, where G is regular open and N is nowhere dense.*

Proof. Let $G = H^{-\prime-\prime}$ and $N = G - H$. Then G is regular open, N is nowhere dense, and $H = G - N$. We have $\bar{N} \subset \bar{G} - H$. Therefore $G - \bar{N} \supset G - (\bar{G} - H) = G \cap H = H$. Also, $H = G - N \supset G - \bar{N}$. Hence $H = G - \bar{N}$. $\square$

Theorem 4.6. *Any set having the property of Baire can be represented in the form $A = G \triangle P$, where G is a regular open set and P is of first category. This representation is unique in any space in which every non-empty open set is of second category (that is, not of first category).*

Proof. The existence of such a representation follows from Theorem 4.5; in any representation we can always replace the open set by the interior of its closure. To prove uniqueness, suppose $G \triangle P = H \triangle Q$, where G is a regular open set, H is open, and P and Q are of first category. Then $H - \bar{G} \subset H \triangle G = P \triangle Q$. Hence $H - \bar{G}$ is an open set of first category, therefore empty. We have $H \subset \bar{G}$, and therefore $H \subset G^{-\prime-\prime} = G$. Thus in the regular open representation the open set G is maximal. If both G and H are regular open, then each contains the other. Hence $G = H$ and $P = Q$. ☐

Theorem 4.7. *The intersection of any two regular open sets is a regular open set.*

Proof. Let $G = G^{-\prime-\prime}$ and $H = H^{-\prime-\prime}$. Since $G \cap H$ is open, it follows that
$$G \cap H \subset (G \cap H)^{-\prime-\prime} \subset G^{-\prime-\prime} = G.$$
Similarly,
$$G \cap H \subset (G \cap H)^{-\prime-\prime} \subset H^{-\prime-\prime} = H.$$
Therefore $G \cap H = (G \cap H)^{-\prime-\prime}$. ☐

All of the foregoing definitions and theorems apply to a space of any number of dimensions (in fact, the proofs are valid in any topological space). Comparison of Theorems 4.3 and 3.16 indicates that the class of sets having the property of Baire is analogous to the class of measurable sets, the sets of first category playing the role of nullsets. It should be noted, however, that in Theorems 4.4 and 3.15 the roles of F_σ and G_δ are interchanged. Moreover, Theorem 4.1 has no analogue for measurable sets; the best one can say is that a measurable set differs from some open (or closed) set by a set of arbitrarily small measure. However, both classes include the Borel sets, and each is invariant under translation. Pursuing the analogy a step further, we have the following theorem, in which $x + A$ denotes the set A translated by x. For simplicity, we confine attention to the 1-dimensional case.

Theorem 4.8. *For any linear set A of second category having the property of Baire, and for any measurable set A with $m(A) > 0$, there exists a positive number δ such that $(x + A) \cap A \neq \emptyset$ whenever $|x| < \delta$.*

Proof. In the first case, let $A = G \triangle P$. Since G is non-empty, it contains an interval I. For any x, we have
$$(x + A) \cap A \supset [(x + I) \cap I] - [P \cup (x + P)].$$

If $|x| < |I|$, the right member represents an interval minus a set of first category; it is therefore non-empty. Hence we may take $\delta = |I|$.

In the second case, let F be a bounded closed subset of A with $m(F) > 0$ (Theorem 3.18). Enclose F in a bounded open set G with $m(G) < (4/3)\,m(F)$. G is the union of a sequence of mutually disjoint intervals. For at least one of these, say I, we must have $m(F \cap I) > (3/4)\,m(I)$. Take $\delta = m(I)/2$. If $|x| < \delta$, then $(x + I) \cup I$ is an interval of length less than $(3/2)\,m(I)$ that contains both $F \cap I$ and $x + (F \cap I)$. These sets cannot be disjoint, because $m(x + (F \cap I)) = m(F \cap I) > (3/4)\,m(I)$. Since $(x + A) \cap A \supset [x + (F \cap I)] \cap [F \cap I]$, it follows that the left member is non-empty. ☐

5. Non-Measurable Sets

Up to now, we have given no indication that the class of measurable sets, or of sets having the property of Baire, does not include all subsets of the line. We know that any set obtained as the result of countably many applications of union, intersection, or complementation, starting from a countable family of closed, open, or nullsets, will be measurable. It can also be shown that any analytic set is measurable. (An *analytic set* is one that can be represented as the continuous image of a Borel set.) According to a result of Gödel [18, p. 388], the hypothesis that there exists a non-measurable set that can be represented as the continuous image of the complement of some analytic set is consistent with the axioms of set theory, provided these axioms are consistent among themselves. No actual example of a non-measurable set that admits such a representation is known (but see [40, p. 17]). Nevertheless, with the aid of the axiom of choice it is easy to show that non-measurable sets exist. We shall consider several such constructions.

The oldest and simplest construction is due to Vitali (1905) [18, p. 59]. Let Q denote the set of rational numbers, considered as a subgroup of the additive group of real numbers. The cosets of Q constitute a partition of the line into an uncountable family of disjoint sets, each congruent to Q under translation. By the axiom of choice, there exists a set V having one and only one element in common with each of these cosets. Let us call any such set a *Vitali set*. The countable family of sets of the form $r + V$ $(r \in Q)$ covers the line. It follows from Theorem 3.19 that V cannot be a nullset. By Theorem 4.8, if V is measurable there exists a number $\delta > 0$ such that $(x + V) \cap V \neq \emptyset$ whenever $|x| < \delta$. But if x is rational and $x \neq 0$, then $(x + V) \cap V = \emptyset$, a contradiction. Hence V cannot be measurable.

Exactly similar reasoning shows that no Vitali set V has the property of Baire. V cannot be of first category, since the sets $r + V$ $(r \in Q)$ cover the line. Then, just as above, Theorem 4.8 implies that V cannot have the property of Baire.

Let $V = A \cup B$ be a partition of a Vitali set V into a set A of first category and a set B of measure zero (Corollary 1.7). Then A is non-measurable but has the property of Baire, while B is measurable but lacks the property of Baire. Thus, neither of these two classes includes the other.

An entirely different construction leading to a non-measurable set is due to F. Bernstein (1908) [18, p. 422]. It is based on the possibility of well ordering a set of power c. First we need

Lemma 5.1. *Any uncountable G_δ subset of R contains a nowhere dense closed set C of measure zero that can be mapped continuously onto $[0, 1]$.*

Proof. Let $E = \bigcap G_n$, G_n open, be an uncountable G_δ set. Let F denote the set of all condensation points of E that belong to E, that is, all points x in E such that every neighborhood of x contains uncountably many points of E. F is non-empty; otherwise, the family of intervals that have rational endpoints and contain only countably many points of E would cover E, and E would be countable. Similar reasoning shows that F has no isolated points. Let $I(0)$ and $I(1)$ be two disjoint closed intervals of length at most $1/3$ whose interiors meet F and whose union is contained in G_1. Proceeding inductively, if 2^n disjoint closed intervals $I(i_1, ..., i_n)$ ($i_k = 0$ or 1) whose interiors all meet F and whose union is contained in G_n have been defined, let $I(i_1, ..., i_{n+1})$ ($i_{n+1} = 0$ or 1) be disjoint closed intervals of length at most $1/3^{n+1}$ contained in $G_{n+1} \cap I(i_1, ..., i_n)$ whose interiors meet F. From the fact that F has no isolated points and that $E \subset G_{n+1}$ it is clear that such intervals exist. Thus a family of intervals $I(i_1, ..., i_n)$ having the stated properties can be defined. Let

$$C = \bigcap_n \ \bigcup_{i_1 ... i_n} \ I(i_1, ..., i_n).$$

Then C is a closed nowhere dense subset of E. C has measure zero for the same reason as the Cantor set. (In fact, C is homeomorphic to the Cantor set.) For each x in C there is a unique sequence $\{i_n\}$, $i_n = 0$ or 1, such that $x \in I(i_1, ..., i_n)$ for every n, and every such sequence corresponds to some point of C. Let $f(x)$ be the real number having the binary development. $i_1 i_2 i_3 ....$ Then f maps C onto $[0, 1]$. Hence C has power c. f is continuous because $|f(x) - f(x')| \leq 1/2^n$ when x and x' both belong to $C \cap I(i_1, ..., i_n)$. ☐

Lemma 5.2. *The class of uncountable closed subsets of R has power c.*

Proof. The class of open intervals with rational endpoints is countable, and every open set is the union of some subclass. Hence there are at most c open sets, and therefore (by complementation) at most c closed sets. On the other hand, there are at least c uncountable closed sets, since there are that many closed intervals. Hence there are exactly c uncountable closed subsets of the line. ☐

Theorem 5.3 (F. Bernstein). *There exists a set B of real numbers such that both B and B' meet every uncountable closed subset of the line.*

By the well ordering principle and Lemma 5.2, the class $\mathscr{F}$ of uncountable closed subsets of the line can be indexed by the ordinal numbers less than ω_c, where ω_c is the first ordinal having c predecessors, say $\mathscr{F} = \{F_\alpha : \alpha < \omega_c\}$. We may assume that R, and therefore each member of $\mathscr{F}$, has been well ordered. Note that each member of $\mathscr{F}$ has power c, by Lemma 5.1, since any closed set is a G_δ. Let p_1 and q_1 be the first two members of F_1. Let p_2 and q_2 be the first two members of F_2 different from both p_1 and q_1. If $1 < \alpha < \omega_c$ and if p_β and q_β have been defined for all $\beta < \alpha$, let p_α and q_α be the first two elements of $F_\alpha - \bigcup_{\beta < \alpha} \{p_\beta, q_\beta\}$. This set is non-empty (it has power c) for each α, and so p_α and q_α are defined for all $\alpha < \omega_c$. Put $B = \{p_\alpha : \alpha < \omega_c\}$. Since $p_\alpha \in B \cap F_\alpha$ and $q_\alpha \in B' \cap F_\alpha$ for each $\alpha < \omega_c$, the set B has the property that both it and its complement meet every uncountable closed set. Let us call any set with this property a *Bernstein set*.

Theorem 5.4. *Any Bernstein set B is non-measurable and lacks the property of Baire. Indeed, every measurable subset of either B or B' is a nullset, and any subset of B or B' that has the property of Baire is of first category.*

Proof. Let A be any measurable subset of B. Any closed set F contained in A must be countable (since every uncountable closed set meets B'), hence $m(F) = 0$. Therefore $m(A) = 0$, by Theorem 3.18. Similarly, if A is a subset of B having the property of Baire, then $A = E \cup P$, where E is G_δ and P is of first category. The set E must be countable, since every uncountable G_δ set contains an uncountable closed set, by Lemma 5.1, and therefore meets B'. Hence A is of first category. The same reasoning applies to B'. ☐

Theorem 5.5. *Any set with positive outer measure has a non-measurable subset. Any set of second category has a subset that lacks the property of Baire.*

Proof. If A has positive outer measure and B is a Bernstein set, Theorem 5.4 shows that the subsets $A \cap B$ and $A \cap B'$ cannot both be measurable. If A is of second category, these two subsets cannot both have the property of Baire. ☐

The fact that every set of positive outer measure has a non-measurable subset was first proved by Rademacher [30] by an entirely different method.

The non-measurability of Vitali's set depended on group-theoretic properties of Lebesgue measure (invariance under translation), that of Bernstein's set depended on topological properties (Theorem 3.18). However, there is an even more fundamental reason, of a purely set-theoretic nature, why (under certain hypotheses) a nontrivial measure cannot be defined for all subsets of a set X. This is the content of a famous theorem of Ulam (1930) [39]. This theorem does not refer directly to measures on the line, but to measures in an abstract set X of restricted cardinality. In the simplest case, it refers to measures in a set of power $\aleph_1$. To say that X has power $\aleph_1$ means that X can be well ordered in such a way that each element is preceded by only countably many elements, that is, the elements of X can be put in one-to-one correspondence with the ordinal numbers less than the first uncountable ordinal.

Theorem 5.6 (Ulam). *A finite measure μ defined for all subsets of a set X of power $\aleph_1$ vanishes identically if it is equal to zero for every one-element subset.*

Proof. By hypothesis, there exists a well ordering of X such that for each y in X the set $\{x : x < y\}$ is countable. Let $f(x, y)$ be a one-to-one mapping of this set onto a subset of the positive integers. Then f is an integer-valued function defined for all pairs (x, y) of elements of X for which $x < y$. It has the property

(1) $$ x < x' < y \quad \text{implies} \quad f(x, y) \neq f(x', y). $$

For each x in X and each positive integer n, define

$$ F_x^n = \{y : x < y, \, f(x, y) = n\} \, . $$

We may picture these sets as arranged in an array

$$
\begin{array}{llll}
F_{x_1}^1 & F_{x_2}^1 & \dots & F_x^1 \dots \\
F_{x_1}^2 & F_{x_2}^2 & \dots & F_x^2 \dots \\
\multicolumn{4}{c}{\dots\dots\dots\dots\dots} \\
F_{x_1}^n & F_{x_2}^n & \dots & F_x^n \dots \\
\multicolumn{4}{c}{\dots\dots\dots\dots\dots}
\end{array}
$$

with $\aleph_0$ rows and $\aleph_1$ columns. This array has the following properties:
(2) The sets in any row are mutually disjoint.

(3) The union of the sets in any column is equal to X minus a countable set.

To verify (2), suppose $y \in F_x^n \cap F_{x'}^n$, for some n and some y, x, and x', with $x \leq x'$. Then $x < y$, $x' < y$, and $f(x, y) = f(x', y) = n$. Hence $x = x'$, by (1). Therefore, for any fixed n, the sets F_x^n ($x \in X$) are disjoint.

To verify (3), observe that if $x < y$, then y belongs to one of the sets F_x^n, namely, that one for which $n = f(x, y)$. Hence the union of the sets F_x^n ($n = 1, 2, \dots$) differs from X by the countable set $\{y : y \leq x\}$.

By (2), in any row there can be at most countably many sets for which $\mu(F_x^n) > 0$ (since $\mu(X)$ is finite). Therefore there can be at most countably many such sets in the whole array. Since there are uncountably many columns, it follows that there exists an element x in X such that $\mu(F_x^n) = 0$ for every n. The union of the sets in this column has measure zero, and the complementary countable set also has measure zero. Therefore $\mu(X) = 0$, and so μ is identically zero. $\square$

Ulam established this result not only for sets X of power $\aleph_1$, but also for some sets of higher cardinality. A limit cardinal is said to be *weakly inaccessible* if (i) it is greater than $\aleph_0$, and (ii) it cannot be represented as a sum of fewer smaller cardinals. It is called *inaccessible* if, in addition, (iii) it exceeds the number of subsets of any set of smaller cardinality. It is easy to see that c is not inaccessible ((iii) fails). If inaccessible cardinals exist, even the smallest ones must be very large. By a continuation of the above reasoning, Ulam showed that in Theorem 5.6 it is sufficient to assume that no cardinal less than or equal to that of X is weakly inaccessible. Neither Theorem 5.6 nor this generalization can be applied to measures on the line unless we make some hypothesis about c. If we assume the *continuum hypothesis* (which asserts that $c = \aleph_1$), or at least if we assume that no cardinal less than or equal to c is weakly inaccessible, then we can infer the following

Proposition 5.7. *A finite measure defined for all subsets of a set of power c vanishes identically if it is zero for points.*

This proposition carries with it a remarkable generalization. In addition to the results mentioned above, Ulam showed that if a set X admits a finite measure μ such that $\mu(X) > 0$ and $\mu(\{x\}) = 0$ for each $x \in X$, and if Proposition 5.7 is true, then X admits a *two-valued* measure (taking only the values 0 and 1) having the same properties. According to a theorem of Hanf and Tarski [32, p. 313], such a measure is possible only if the cardinal number of X is enormously large, in fact, the cardinal number of X must be preceded by an equal number of inaccessible cardinals!

It should be pointed out that the non-measurability of a set does not mean that no measure can be defined for it. In fact, it can be shown that any subset of R is included in the domain of some extension of Lebesgue measure. However, Theorem 5.6 shows that if $c = \aleph_1$, then no extension of Lebesgue measure can be defined for every member of the array $\{F_x^n\}$. Even more remarkably, Banach [1] has shown that the continuum hypothesis implies the existence of a *countable* family of sets that has this property. But it should not be forgotten that unless we assume the continuum hypothesis, or make some special hypothesis concerning c, neither Proposition 5.7 nor the impossibility of extending Lebesgue measure to all subsets of R has yet been proved.

6. The Banach-Mazur Game

Around 1928, the Polish mathematician S. Mazur invented the following mathematical "game." Player (A) is "dealt" an arbitrary subset A of a closed interval I_0. The complementary set $B = I_0 - A$ is dealt to player (B). The game $\langle A, B \rangle$ is played as follows: (A) chooses arbitrarily a closed interval $I_1 \subset I_0$; then (B) chooses a closed interval $I_2 \subset I_1$; then (A) chooses a closed interval $I_3 \subset I_2$; and so on, alternately. Together the players determine a nested sequence of closed intervals I_n, (A) choosing those with odd index, (B) those with even index. If the set $\bigcap I_n$ has at least one point in common with A, then (A) wins; otherwise, (B) wins.

The question is: can one of the players, by choosing his intervals judiciously, insure that he will win no matter how his opponent plays? Anyone familiar with the proof of the Baire category theorem can hardly fail to notice that in case the set A is of first category, there is a simple strategy by which (B) can insure that he will win. If $A = \bigcup A_n$, A_n nowhere dense, (B) has only to choose $I_{2n} \subset I_{2n-1} - A_n$, for each n. Then no matter how (A) plays, (B) will win. Mazur conjectured that *only* when A is of first category can the second player be sure to win. Banach (unpublished) proved this conjecture to be true [40, p. 23], [9].

To say precisely what it means for one of the players to be sure to win, we need to understand what is meant by a "strategy." A strategy for either player is a rule that specifies what move he will make in every possible situation. At his n-th move, (B) knows which intervals $I_0, I_1, \ldots, I_{2n-1}$ have been chosen in the previous moves, and he knows the sets A and B, but that is all. From this information, his strategy must tell him which interval to choose for I_{2n}. Thus, a strategy for (B) is a sequence of closed-interval-valued functions $f_n(I_0, I_1, \ldots, I_{2n-1})$. The rules of the game demand that

$$(1) \qquad f_n(I_0, I_1, \ldots, I_{2n-1}) \subset I_{2n-1} \qquad (n = 1, 2, \ldots).$$

The function f_n must be defined at least for all intervals that satisfy the conditions

$$(2) \qquad I_0 \supset I_1 \supset I_2 \supset \cdots \supset I_{2n-1}$$

and

(3) $$I_{2i} = f_i(I_0, I_1, \ldots, I_{2i-1}) \quad (i = 1, 2, \ldots, n-1).$$

For this to be a winning strategy for (B), it is necessary and sufficient that $\bigcap I_n \subset B$ for every sequence I_n that satisfies (2) and (3) for every n.

Theorem 6.1. *There exists a strategy by which (B) can be sure to win if and only if A is of first category.*

Proof. Let $f_1, f_2, \ldots$ be a winning strategy for (B). Let I^0 denote the interior of any interval I. Given f_1, it is possible to define a sequence of closed intervals J_i $(i = 1, 2, \ldots)$ contained in I_0^0 such that (i) the intervals $K_i = f_1(I_0, J_i)$ are disjoint, and (ii) the union of their interiors is dense in I_0. One way to do this is as follows. Let S be a sequence consisting of all closed intervals that have rational endpoints and are contained in I_0^0. Let J_1 be the first term of S. Having defined $J_1, \ldots, J_i$, let J_{i+1} be the first term of S contained in $I_0 - K_1 - K_2 - \cdots - K_i$. It is easy to verify, using (1), that this construction defines inductively a sequence J_i having the required properties.

Similarly, for each i, let J_{ij} $(j = 1, 2, \ldots)$ be a sequence of closed intervals contained in K_i^0 such that the intervals $K_{ij} = f_2(I_0, J_i, K_i, J_{ij})$ are disjoint and the union of their interiors is dense in K_i. Then the union of all the intervals K_{ij}^0 is dense in I_0.

Proceeding inductively, we can define two families of closed intervals $J_{i_1 \ldots i_n}$ and $K_{i_1 \ldots i_n}$, where n and each of the indices i_k range over all positive integers, such that the following conditions are satisfied:

(4) $$K_{i_1 \ldots i_n} = f_n(I_0, J_{i_1}, K_{i_1}, J_{i_1 i_2}, K_{i_1 i_2}, \ldots, J_{i_1 \ldots i_n}),$$

(5) $$J_{i_1 \ldots i_{n+1}} \subset K_{i_1 \ldots i_n}^0.$$

(6) For each n, the intervals $K_{i_1 \ldots i_n}$ are disjoint, and the union of their interiors is dense in I_0.

Now consider an arbitrary sequence of positive integers i_n, and define

(7) $$I_{2n-1} = J_{i_1 \ldots i_n}, \quad I_{2n} = K_{i_1 \ldots i_n} \quad (n = 1, 2, \ldots).$$

From (4) and (5) it follows that conditions (2) and (3) are satisfied for all n; hence the nested sequence I_n is a possible play of the game consistent with the given strategy for (B). By hypothesis, the set $\bigcap I_n$ must be contained in B.

For each n, define $G_n = \bigcup_{i_1 \ldots i_n} K_{i_1 \ldots i_n}^0$. Let $E = \bigcap G_n$. Then for each x in E there is a unique sequence $i_1, i_2, \ldots$ such that $x \in K_{i_1 \ldots i_n}$ for every n. If this sequence is used to define (7), then $x \in \bigcap I_n \subset B$. This shows that $E \subset B$. Consequently,

$$A = I_0 - B \subset I_0 - E = \bigcup_n (I_0 - G_n).$$

Since (6) implies that each of the sets $I_0 - G_n$ is nowhere dense, A must be of first category. ☐

This theorem gives new insight into the sense in which a set of first category is small; it is a set on which even the first player is bound to lose, unless his opponent fails to take advantage of the situation.

Theorem 6.2. *There exists a strategy by which (A) can be sure to win if and only if $I_1 \cap B$ is of first category for some interval $I_1 \subset I_0$.*

Proof. If such an interval exists, (A) can start by choosing it for I_1. Then, by an obvious strategy, he can insure that $\bigcap I_n$ is disjoint to B. Since the intersection is non-empty, this is a winning strategy for (A). On the other hand, if (A) has a winning strategy he can always modify it so as to insure that the intersection of the intervals I_n will consist of just one point of A. (For instance, this will be insured if he always chooses I_{2n+1} as if I_{2n} had been a subinterval half as long.) This defines a winning strategy for the second player in the game $\langle I_1 \cap B, I_1 \cap A \rangle$. By Theorem 6.1, such a strategy can exist only if $I_1 \cap B$ is of first category. ☐

Theorem 6.3. *If the set A has the property of Baire, then (B) or (A) possesses a winning strategy according as A is of first or second category.*

Proof. Let $A = G \triangle P$, where G is open and P is of first category. If G is empty, then (B) has a winning strategy, by Theorem 6.1. If G is not empty, (A) has only to choose $I_1 \subset G$ to insure that he will be able to win. ☐

A set E is said to be of *first category at the point* x if there exists a neighborhood U of x such that $U \cap E$ is of first category. Otherwise, E is said to be of *second category at the point* x. These notions are analogous to the metric notion of density discussed in Chapter 3. The set G of points at which A' is of first category is open. If A has the property of Baire, G may be regarded as the category analogue of the set $\phi(A)$ considered in Theorem 3.21; it is the largest open set that differs from A by a set of first category. Hence G is the same as the regular open set that appears in Theorem 4.6. The fact that G differs from A by a set of first category is analogous to the Lebesgue density theorem.

By Theorems 6.1 and 6.2, one of the players possesses a winning strategy if and only if A is of first category or B is of first category at some point. By Theorem 6.3, one of these alternatives holds whenever A has the property of Baire. Is it possible that neither may hold? Yes! Let A be the intersection of I_0 with a Bernstein set. Then neither A nor B contains an uncountable G_δ set (Lemma 5.1). Consequently, for any interval $I \subset I_0$, neither of the sets $A \cap I$ or $B \cap I$ is of first category. (For if one is of first category, the other is a set of second category having the property of Baire. By Theorem 4.4, any such set contains an uncountable G_δ set.) Consequently, this game $\langle A, B \rangle$ is not determined in favor of either player.

The possibility of indeterminateness makes the Banach-Mazur game particulary interesting for the general theory of games. It also raises some interesting questions. If a game is determined in favor of one of the players, should it be called a game of "skill"? If neither player can control the outcome, is the outcome a matter of "chance"? What does "chance" mean in this connection?

There is another version of the Banach-Mazur game, in which the players alternately choose successive blocks of digits (of arbitrary finite length) in the decimal (or binary) development of a number. If the number so defined belongs to A, (A) wins; otherwise, (B) wins. In effect, this is the same as the game with intervals, except that now all the intervals are required to be decimal intervals. Any winning strategy for the original game can easily be modified so as to satisfy this condition, and Theorems 6.1, 6.2, and 6.3 remain valid. However, if the blocks are all required to be of length 1, that is, if (A) and (B) alternately choose successive digits in the development of a number, then we have an entirely different game, a game which was first studied by Gale and Stewart [8]. The conditions under which one of the players now has a winning strategy are still not completely understood. It is not known, for instance, whether this game is determined in favor of one or the other player whenever A is a Borel set. Recent results suggest that the answer may depend upon what set-theoretic axioms one assumes [22, p. 75].

7. Functions of First Class

Let f be a real-valued function on R. For any interval I, the quantity

$$\omega(I) = \sup_{x \in I} f(x) - \inf_{x \in I} f(x)$$

is called the *oscillation of f on I*. For any fixed x, the function $\omega((x - \delta, x + \delta))$ decreases with δ and approaches a limit

$$\omega(x) = \lim_{\delta \to 0} \omega((x - \delta, x + \delta)),$$

called the *oscillation of f at x*. $\omega(x)$ is an extended real-valued function on R. Evidently, $\omega(x_0) = 0$ if and only if f is continuous at x_0. When it is not zero, $\omega(x_0)$ is a measure of the size of the discontinuity of f at x_0.

If $\omega(x_0) < \varepsilon$, then $\omega(x) < \varepsilon$ for all x in a neighborhood of x_0. Hence the set $\{x : \omega(x) < \varepsilon\}$ is open. The set D of all points at which f is discontinuous can be represented in the form

$$D = \bigcup_{n=1}^{\infty} \{x : \omega(x) \geq 1/n\},$$

hence D is always an F_σ set. Thus

Theorem 7.1. *If f is a real-valued function on R, then the set of points of discontinuity of f is an F_σ.*

This theorem admits the following converse:

Theorem 7.2. *For any F_σ set E there exists a bounded function f having E for its set of points of discontinuity.*

Proof. Let $E = \bigcup F_n$, where F_n is closed. We may assume that $F_n \subset F_{n+1}$ for all n. Let A_n denote the set of rational points interior to F_n. For any set A, the function χ_A defined by

$$\chi_A(x) = \begin{cases} 1 & \text{when} \quad x \in A \\ 0 & \text{when} \quad x \notin A \end{cases}$$

is called the *indicator function* (or *characteristic function*) of A. The function $f_n = \chi_{F_n} - \chi_{A_n} = \chi_{F_n - A_n}$ has oscillation equal to 1 at each point of F_n, and equal to 0 elsewhere. Let $\{a_n\}$ be a sequence of positive numbers

31

such that $a_n > \sum_{i>n} a_i$ for every n. (For instance, let $a_n = 1/n!$.) Then the series $\sum_{n=1}^{\infty} a_n f_n(x)$ converges uniformly on R to a bounded function f. f is continuous at any point where all of the terms are continuous, hence at each point of $R - E$. On the other hand, at each point of $F_n - F_{n-1}$ the oscillation of f is at least equal to $a_n - \sum_{i>n} a_i$. Hence the set of points of discontinuity of f is exactly E. ⬚

A function f is said to be of the *first class* (of Baire) if it can be represented as the limit of an everywhere convergent sequence of continuous functions. Such a function need not be continuous, as simple examples show. For instance, the functions $f_n(x) = \max(0, 1 - n|x|)$ are continuous and the sequence converges pointwise to the discontinuous function $f(x) = 1$ or 0 according as $x = 0$ or $x \neq 0$. However, the following theorem shows that a function of first class cannot be everywhere discontinuous. It is known as *Baire's theorem on functions of first class.* (More exactly, it is a part of Baire's theorem.) It was in this connection that Baire originally introduced the notion of category.

Theorem 7.3. *If f can be represented as the limit of an everywhere convergent sequence of continuous functions, then f is continuous except at a set of points of first category.*

This should be compared with the well-known theorem that the limit of a *uniformly* convergent sequence of continuous functions is everywhere continuous.

Proof. It suffices to show that, for each $\varepsilon > 0$, the set $F = \{x : \omega(x) \geq 5\varepsilon\}$ is nowhere dense. Let $f(x) = \lim f_n(x)$, f_n continuous, and define

$$E_n = \bigcap_{i,j \geq n} \{x : |f_i(x) - f_j(x)| \leq \varepsilon\} \qquad (n = 1, 2, \ldots).$$

Then E_n is closed, $E_n \subset E_{n+1}$, and $\bigcup E_n$ is the whole line. Consider any closed interval I. Since $I = \bigcup (E_n \cap I)$, the sets $E_n \cap I$ cannot all be nowhere dense. Hence, for some positive integer n, $E_n \cap I$ contains an open interval J. We have $|f_i(x) - f_j(x)| \leq \varepsilon$ for all x in J; $i, j \geq n$. Putting $j = n$ and letting $i \to \infty$, it follows that $|f(x) - f_n(x)| \leq \varepsilon$ for all x in J. For any x_0 in J there is a neighborhood $I(x_0) \subset J$ such that $|f_n(x) - f_n(x_0)| \leq \varepsilon$ for all x in $I(x_0)$. Hence $|f(x) - f_n(x_0)| \leq 2\varepsilon$ for all x in $I(x_0)$. Therefore $\omega(x_0) \leq 4\varepsilon$, and so no point of J belongs to F. Thus for every closed interval I there is an open interval $J \subset I - F$. This shows that F is nowhere dense. ⬚

The reasoning just given can be used to prove more. With only slight changes in wording, it applies when f and all of the functions f_n are restricted to an arbitrary perfect set P. In this case the notion of category must be interpreted relative to P. The Baire category theorem remains true: if an open interval I meets P, then no countable union of sets nowhere dense relative to P can be equal to $I \cap P$. Thus if f is any function of first class and P is any perfect set, then the restriction of f to P is continuous at all points of P except a set of first category relative to P. Conversely, Baire showed that any such function is of first class. (For an elementary proof, see [4, Note II].) We shall not prove this, but merely note that a simple example shows that the converse of Theorem 7.3 is false. Let $f(x) = 0$ at all points not in the Cantor set C, $f(x) = 1/2$ at the endpoints of each of the open intervals deleted in the construction of C, and $f(x) = 1$ at all other points of C. f is continuous except at a set of points of first category, namely, at every point of C'. But the restriction of f to C is discontinuous at every point of C, hence f is not of first class.

It is easy enough to formulate a necessary and sufficient condition for the conclusion of Theorem 7.3, namely,

Theorem 7.4. *Let f be a real-valued function on R. The set of points of discontinuity of f is of first category if and only if f is continuous at a dense set of points.*

This is an immediate consequence of Theorem 7.1 and the fact that an F_σ set is of first category if and only if its complement is dense.

Theorem 7.3 is an extremely useful result. To illustrate how it serves to answer several natural questions, we mention two examples.

It is well known that a trigonometric series may converge pointwise to a discontinuous function. How discontinuous can the sum function be? Can the sum of an everywhere convergent trigonometric series be everywhere discontinuous? Theorem 7.3 shows at once that it cannot.

Again, it is well known that the derivative of an everywhere differentiable function f need not be everywhere continuous. A familiar example is the function

$$f(x) = x^2 \sin(1/x), \quad f(0) = 0.$$

Can the derivative of an everywhere differentiable function be everywhere discontinuous? Theorem 7.3 answers the question, since

$$f'(x) = \lim_{n \to \infty} \frac{f(x + 1/n) - f(x)}{1/n}$$

is a function of first class when it is everywhere defined and finite.

Having found conditions under which the set D of points of discontinuity of a function is of first category, it is natural to inquire under what conditions D is a nullset. One answer is provided by the following well-known

Theorem 7.5. *In order that a function f be Riemann-integrable on every finite interval it is necessary and sufficient that f be bounded on every finite interval and that its set of points of discontinuity be a nullset.*

For a given function f, bounded on I, let $F(I)$ be the greatest lower bound of all sums of the form

$$\sum_{i=1}^{n} \omega(I_i)|I_i|,$$

where $\{I_1, \ldots, I_n\}$ is any subdivision of I, that is, any finite set of non-overlapping closed intervals whose union is I. $F(I)$ is the difference between the upper and lower integrals of f on I; the equation $F(I) = 0$ expresses the condition that f be Riemann-integrable on I. It is easy to verify that if $\{I_1, \ldots, I_n\}$ is any subdivision of I, then $F(I) = \sum_1^n F(I_i)$. This property of F is all that is needed to prove the following

Lemma 7.6. *If $\omega(x) < \varepsilon$ for each x in I, then $F(I) < \varepsilon|I|$.*

Proof. Suppose the contrary. Then $F(I) \geq \varepsilon|I|$, and so $F(I_1) \geq \varepsilon|I|/2$ for at least one of the intervals I_1 obtained by bisecting I. Similarly, $F(I_2) \geq \varepsilon|I_1|/2$ for at least one of the intervals I_2 obtained by bisecting I_1. By repeated bisection we obtain a nested sequence of closed intervals I_n such that $F(I_n) \geq \varepsilon|I|/2^n$ $(n = 1, 2, \ldots)$. These intersect in a point x of I. By hypothesis, $\omega(x) < \varepsilon$ and therefore $\omega(J) < \varepsilon$ for some open interval J containing x. Choose n so that $I_n \subset J$. Then

$$F(I_n) \leq \omega(I_n)|I_n| \leq \omega(J)|I|/2^n < \varepsilon|I|/2^n \leq F(I_n),$$

a contradiction. ☐

Corollary 7.7. *Any continuous function on a closed interval is integrable.*

It may be noted that the above proof of this fact did not involve the notion of uniform continuity.

Now, to prove Theorem 7.5, assume first that f is integrable on I. Then for any positive integer k, I can be divided into intervals $I_1, \ldots, I_n$ such that

$$\sum_{i=1}^{n} \omega(I_i)|I_i| < 1/k^2.$$

Let $\sum'$ denote the sum over those intervals I_i for which $\omega(x) \geq 1/k$ at some interior point. Then

$$1/k^2 > \sum' \omega(I_i)|I_i| \geq (1/k) \sum'|I_i|.$$

Therefore $\sum'|I_i| < 1/k$. The set

$$F_k = \{x \in I : \omega(x) \geq 1/k\}$$

is entirely covered by these intervals, except perhaps for a finite number of points (endpoints of intervals of the subdivision). Therefore $m(F_k) < 1/k$. If D is the set of points of discontinuity of f, then $D \cap I$ is the union of the increasing sequence F_k, and we have

$$m(D \cap I) = \lim_{k \to \infty} m(F_k) = 0.$$

If f is integrable on every finite interval, it follows that D is a nullset.

Conversely, suppose D is a nullset and that f is bounded on I, with upper and lower bounds M and m, respectively. For any $\varepsilon > 0$, choose k so that $(M - m) + |I| < k\varepsilon$. Since F_k is a bounded closed nullset, it is possible to cover F_k with a finite number of disjoint open intervals the sum of whose lengths is less than $1/k$. The endpoints of these intervals that belong to I determine a subdivision of I into nonoverlapping intervals I_i and J_j such that $\sum |I_i| < 1/k$ and $\omega(x) < 1/k$ on each of the intervals J_j. Hence, by Lemma 7.6,

$$F(I) = \sum F(I_i) + \sum F(J_j) \leqq (M - m)\sum |I_i| + \sum (1/k)|J_j|$$
$$\leqq (M - m)/k + |I|/k < \varepsilon .$$

Consequently, f is Riemann-integrable on I.

To round out this discussion of points of discontinuity, one may ask whether there is a natural class of functions that is characterized by having only countably many discontinuities. One answer is provided by

Theorem 7.8. *The set of points of discontinuity of any monotone function f is countable. Any countable set is the set of points of discontinuity of some monotone function.*

Proof. If f is monotone, there can be at most $|f(b) - f(a)|/\varepsilon$ points in (a, b) where $\omega(x) \geqq \varepsilon$. Hence the set of points of discontinuity of f is countable. On the other hand, let $\{x_i\}$ be any countable set, and let $\sum \varepsilon_i$ be a convergent series of positive real numbers. The function $f(x) = \sum_{x_i \leqq x} \varepsilon_i$ is a monotone bounded function. It has the property that $\omega(x_i) = \varepsilon_i$ for each i, and $\omega(x) = 0$ for all x not in the sequence x_i. $\quad\square$

This should be compared with the much deeper theorem, due to Lebesgue, that any monotone function is differentiable (has a finite derivative) except at a set of points of measure zero [31, p. 5].

8. The Theorems of Lusin and Egoroff

A real-valued function f on R is called *measurable* if $f^{-1}(U)$ is measurable for every open set U in R. f is said to have the *property of Baire* if $f^{-1}(U)$ has the property of Baire for every open set U in R. In either definition, U may be restricted to some base, or allowed to run over all Borel sets. The indicator function χ_E of a set $E \subset R$ is measurable if and only if E is measurable; χ_E has the property of Baire if and only if E does.

If E has the property of Baire, then $E = G \triangle P = F \triangle Q$, where G is open, F is closed, and P and Q are of first category. The set $E - (P \cup Q)$ $= G - (P \cup Q) = F - (P \cup Q)$ is both closed and open relative to $R - (P \cup Q)$, hence the restriction of χ_E to the complement of $P \cup Q$ is continuous. More generally, continuity and the property of Baire are related as follows [18, p. 306].

Theorem 8.1. *A real-valued function f on R has the property of Baire if and only if there exists a set P of first category such that the restriction of f to $R - P$ is continuous.*

Proof. Let $U_1, U_2, \ldots$ be a countable base for the topology of R, for example, the open intervals with rational endpoints. If f has the property of Baire, then $f^{-1}(U_i) = G_i \triangle P_i$, where G_i is open and P_i is of first category. Put $P = \bigcup_1^\infty P_i$. Then P is of first category. The restriction g of f to $R - P$ is continuous, since $g^{-1}(U_i) = f^{-1}(U_i) - P = (G_i \triangle P_i) - P$ $= G_i - P$ is open relative to $R - P$ for each i, and therefore so is $g^{-1}(U)$, for every open set U.

Conversely, if the restriction g of f to the complement of some set P of first category is continuous, then for any open set U, $g^{-1}(U) = G - P$ for some open set G. Since

$$g^{-1}(U) \subset f^{-1}(U) \subset g^{-1}(U) \cup P,$$

we have

$$G - P \subset f^{-1}(U) \subset G \cup P.$$

Therefore $f^{-1}(U) = G \triangle Q$ for some set $Q \subset P$. Thus f has the property of Baire. □

The relation between continuity and measurability is not quite so simple. It is expressed by the following, known as *Lusin's theorem.*

Theorem 8.2 (Lusin). *A real-valued function f on R is measurable if and only if for each $\varepsilon > 0$ there exists a set E with $m(E) < \varepsilon$ such that the restriction of f to $R - E$ is continuous.*

Proof. Let $U_1, U_2, \ldots$ be a countable base for the topology of R. If f is measurable, then for each i there exists a closed set F_i and an open set G_i such that

$$F_i \subset f^{-1}(U_i) \subset G_i \quad \text{and} \quad m(G_i - F_i) < \varepsilon/2^i .$$

Put $E = \bigcup_1^\infty (G_i - F_i)$. Then $m(E) < \varepsilon$. If g denotes the restriction of f to $R - E$, then

$$g^{-1}(U_i) = f^{-1}(U_i) - E = F_i - E = G_i - E .$$

Hence $g^{-1}(U_i)$ is both closed and open relative to $R - E$, and it follows that g is continuous.

Conversely, if f has the stated property there is a sequence of sets E_i with $m(E_i) < 1/i$ such that the restriction f_i of f to $R - E_i$ is continuous. For any open set U there are open sets G_i such that $f_i^{-1}(U) = G_i - E_i$ $(i = 1, 2, \ldots)$. Putting $E = \bigcap_1^\infty E_i$, we have

$$f^{-1}(U) - E = \bigcup_{i=1}^\infty (f^{-1}(U) - E_i) = \bigcup_{i=1}^\infty f_i^{-1}(U) .$$

Consequently,

$$f^{-1}(U) = [f^{-1}(U) \cap E] \cup \bigcup_{i=1}^\infty (G_i - E_i) .$$

All of these sets are measurable, since $m(E) = 0$, and therefore f is a measurable function. $\square$

A measurable function need not be continuous on the complement of a nullset. To see this we construct an example as follows. Let $U_1, U_2, \ldots$ be a base for the topology of R. Since every interval contains a nowhere dense set of positive measure, we can define inductively a disjoint sequence of nowhere dense closed sets N_n such that $m(N_n) > 0$ and $N_{2n} \cup N_{2n-1} \subset U_n$. Put $A = \bigcup_1^\infty N_{2n}$, and let f be the indicator function of A. Since A and $R - A$ have positive measure in every interval, the restriction of f to the complement of any nullset is nowhere continuous.

The following result, known as *Egoroff's theorem,* establishes a relation between convergence and uniform convergence.

Theorem 8.3. *If a sequence of measurable functions f_n converges to f at each point of a set E of finite measure, then for each $\varepsilon > 0$ there is a set $F \subset E$ with $m(F) < \varepsilon$ such that f_n converges to f uniformly on $E - F$.*

Proof. For any two positive integers n and k let

$$E_{n,k} = \bigcup_{i=n}^\infty \{x \in E : |f_i(x) - f(x)| \geq 1/k\} .$$

37

Then $E_{n,k} \supset E_{n+1,k}$ and $\bigcap_{n=1}^{\infty} E_{n,k} = \emptyset$, for each k. Given $\varepsilon > 0$, for each k there is an integer $n(k)$ such that $m(E_{n(k),k}) < \varepsilon/2^k$. Put $F = \bigcup_{k=1}^{\infty} E_{n(k),k}$. Then $m(F) < \varepsilon$. For each k we have $E - F \subset E - E_{n(k),k}$. Therefore $|f_i(x) - f(x)| < 1/k$ for all $i \geq n(k)$ and all $x \in E - F$. Thus f_n converges to f uniformly on $E - F$. $\square$

It is interesting to note that while Lusin's theorem has a very satisfactory category analogue in Theorem 8.1, the corresponding analogue of Egoroff's theorem is false. This is shown by the following example.

Let $\phi(x)$ be the piece-wise linear continuous function defined by $\phi(x) = 2x$ on $[0, 1/2]$, $\phi(x) = 2 - 2x$ on $[1/2, 1]$, and $\phi(x) = 0$ on $R - [0, 1]$. Then $\lim_{n \to \infty} \phi(2^n x) = 0$ for every x in R. Let $\{r_i\}$ be a dense sequence in R, and define $f_n(x) = \sum_{i=1}^{\infty} 2^{-i} \phi(2^n(x - r_i))$. As the sum of a uniformly convergent series of continuous functions, f_n is continuous on R, and $\lim_{n \to \infty} f_n(x) = 0$ for each x in R. If (a, b) is any open interval, then $r_i \in (a, b)$ for some i, and we have $\sup_{a < x < b} f_n(x) \geq 1/2^i$ for all sufficiently large n. This shows that f_n does not converge uniformly on (a, b). Let E be any set on which f_n does converge uniformly. This means that if we let $\alpha_n = \sup_{x \in E} f_n(x)$, then $\alpha_n \to 0$. Because f_n is continuous, α_n is also the supremum of f_n on $\overline{E}$. Hence f_n converges to 0 uniformly on $\overline{E}$. From what we have shown, $\overline{E}$ cannot contain an interval. Therefore any set on which the sequence $\{f_n\}$ converges uniformly is nowhere dense.

9. Metric and Topological Spaces

The usefulness of the notion of category only becomes fully apparent in more general spaces, especially metric spaces. Let us recall the basic definitions.

A *metric space* is a set X together with a *distance function* or *metric* $\varrho(x, y)$ defined for all pairs of points of X and satisfying the following conditions:

(1) $\varrho(x, y) \geqq 0$, $\varrho(x, x) = 0$,

(2) $\varrho(x, y) = \varrho(y, x)$,

(3) $\varrho(x, z) \leqq \varrho(x, y) + \varrho(y, z)$ (triangle inequality),

(4) $\varrho(x, y) = 0$ implies $x = y$.

This notion (due to Fréchet) is a natural abstraction of some of the properties of distance in a Euclidean space of any number of dimensions. Many theorems in analysis become simpler and more intuitive when formulated in terms of a suitable metric.

A sequence $x_1, x_2, \ldots$ of points of a metric space (X, ϱ) is said to *converge to the point* x if $\varrho(x_n, x) \to 0$ as $n \to \infty$. We then write $x_n \to x$. A sequence is *convergent* if it converges to some point of X. The set of points $\{x : \varrho(x_0, x) < r\}$, $r > 0$, is called the *r-neighborhood of* x_0, or the *ball* with center x_0 and radius r.

A set $G \subset X$ is called *open* if for each x in G, G contains some ball with center x. Balls are open sets, and arbitrary unions and finite intersections of open sets are open. Any class $\mathscr{T}$ of subsets of a set X such that $\emptyset$, X, the union of any subclass of $\mathscr{T}$, and the intersection of any finite subclass of $\mathscr{T}$ belongs to $\mathscr{T}$ is called a *topology* in X, and the pair $(X, \mathscr{T})$ is called a *topological space*. A subclass $\mathscr{T}_0 \subset \mathscr{T}$ is a *base* for the topology if each member of $\mathscr{T}$ is the union of some subfamily of $\mathscr{T}_0$. The open subsets of any metric space X constitute a topology in X, but not every topology can be represented in this way.

A metric space is called *separable* if it has a countable dense subset, or, equivalently, a countable base. Two metrics in a set X are *topologically equivalent* if they determine the same topology. Very often it is the topological structure of a metric space that is of primary interest, and the metric is regarded as auxiliary. Any property that is definable in terms of open sets alone is a topological property. For instance, convergence is a topological property, because $x_n \to x$ if and only if every open set that contains x contains all but a finite number of terms of the sequence.

The complement of an open set is called *closed*. In a metric space X, a set F is closed if and only if $\{x_n\} \subset F$, $x_n \to x$ imply $x \in F$. The smallest closed set that contains a set A is called the *closure* of A; it is denoted by $\bar{A}$ or A^-. Similarly, the largest open set contained in A is called the *interior* of A; it is equal to $A'^{-\prime}$. A is a *neighborhood* of x if x belongs to the interior of A. A set A is *dense* (in X) if $\bar{A} = X$, that is, if every non-empty open set contains at least one point of A. A set A is *nowhere dense* if the interior of its closure is empty, that is, if for every non-empty open set G there is a non-empty open set H contained in $G - A$. A set is of *first category* if it can be represented as a countable union of nowhere dense sets; otherwise, it is of *second category*. F_σ sets, G_δ sets, Borel sets, and sets having the property of Baire are defined exactly as before. All of these are topological properties of sets, and the definitions apply to any topological space.

A mapping f of a topological space X into a topological space Y is *continuous at the point* x_0 in X if for every open set V that contains $f(x_0)$ there is a neighborhood U of x_0 such that $f(x) \in V$ for every $x \in U$. A mapping $f : X \to Y$ is *continuous* if it is continuous at each point of X. A one-to-one mapping f of X onto Y is called a *homeomorphism* if both f and f^{-1} are continuous. When such a mapping exists, X and Y are said to be *homeomorphic* or *topologically equivalent*. Two metrics ϱ and σ in a set X are topologically equivalent if and only if the identity mapping of X onto itself is a homeomorphism of (X, ϱ) onto (X, σ). For this it is necessary and sufficient that $\varrho(x_n, x) \to 0$ if and only if $\sigma(x_n, x) \to 0$.

A sequence of points x_n of a metric space (X, ϱ) is called a *Cauchy sequence* if for each $\varepsilon > 0$ there is a positive integer n such that $\varrho(x_i, x_j) < \varepsilon$ for all $i, j \geq n$. Every convergent sequence is Cauchy, but the converse is not generally true. However, there is an important class of spaces in which every Cauchy sequence is convergent. Such a metric space is said to be *complete*. For instance, the real line is complete with respect to the usual metric $|x - y|$.

It is important to realize that completeness is not a topological property, and that the class of Cauchy sequences (unlike the class of convergent sequences) is not preserved under homeomorphism. For instance, the mapping that takes x into arc tan x is a homeomorphism of the line X onto the open interval $Y = (-\pi/2, \pi/2)$. Here X is complete but Y is not. The sequence $y_n = $ arc tan n is Cauchy in Y, but the sequence $x_n = n$ is not Cauchy in X.

A metric space (X, ϱ) is *topologically complete* if it is homeomorphic to some complete space. If f is a homeomorphism of (X, ϱ) onto a complete space (Y, σ), then $\sigma(f(x), f(y))$ is a metric in X topologically equivalent to ϱ. Thus a metric space is topologically complete if and only if it can be remetrized (with a topologically equivalent metric) so as to be complete. An important property of such spaces is that the Baire category theorem still holds.

Theorem 9.1. *If X is a topologically complete metric space, and if A is of first category in X, then $X - A$ is dense in X.*

Proof. Let $A = \bigcup A_n$, where A_n is nowhere dense, let ϱ be a metric with respect to which X is complete, and let S_0 be a non-empty open set. Choose a nested sequence of balls S_n of radius $r_n < 1/n$ such that $\bar{S}_n \subset S_{n-1} - A_n$ ($n \geq 1$). This can be done step by step, taking for S_n a ball with center x_n in $S_{n-1} - \bar{A}_n$ (which is non-empty because $\bar{A}_n$ is nowhere dense) and with sufficiently small radius. Then $\{x_n\}$ is a Cauchy sequence, since

$$\varrho(x_i, x_j) \leq \varrho(x_i, x_n) + \varrho(x_n, x_j) < 2r_n \quad \text{for} \quad i, j \geq n .$$

Hence $x_n \to x$ for some x in X. Since $x_i \in \bar{S}_n$ for $i \geq n$, it follows that $x \in \bigcap \bar{S}_n \subset S_0 - A$. This shows that $X - A$ is dense in X. $\square$

A topological space X is called a *Baire space* if every non-empty open set in X is of second category, or equivalently, if the complement of every set of first category is dense. In a Baire space, the complement of any set of first category is called a *residual set*.

Theorem 9.2. *In a Baire space X, a set E is residual if and only if E contains a dense G_δ subset of X.*

Proof. Suppose $B = \bigcap G_n$, G_n open, is a G_δ subset of E that is dense in X. Then each G_n is dense, and $X - E \subset X - B = \bigcup (X - G_n)$ is of first category. Conversely, if $X - E = \bigcup A_n$, where A_n is nowhere dense, let $B = \bigcap (X - \bar{A}_n)$. Then B is a G_δ set contained in E. Its complement $X - B = \bigcup \bar{A}_n$ is of first category. Since X is a Baire space, it follows that B is dense in X. $\square$

10. Examples of Metric Spaces

Let C, or $C[a, b]$, denote the set of all real-valued continuous functions f on the interval $[a, b]$, and define

$$\varrho(f, g) = \sup_{a \leq x \leq b} |f(x) - g(x)|.$$

It is easy to verify that ϱ is a metric in C; in particular, the triangle inequality follows from the fact that

$$|f(x) - h(x)| \leq |f(x) - g(x)| + |g(x) - h(x)|$$
$$\leq \varrho(f, g) + \varrho(g, h)$$

for all x in $[a, b]$. Convergence in this metric means uniform convergence on $[a, b]$. For this reason, ϱ is called the *uniform metric*.

Let $\{f_n\}$ be any Cauchy sequence in C, say $\varrho(f_i, f_j) \leq \varepsilon$ for all $i, j \geq n(\varepsilon)$. Then

$$|f_i(x) - f_j(x)| \leq \varepsilon \quad \text{for all} \quad i, j \geq n(\varepsilon) \quad \text{and} \quad a \leq x \leq b.$$

Hence, for each x in $[a, b]$, $\{f_n(x)\}$ is a Cauchy sequence of real numbers. It therefore converges to a limit $f(x)$. Letting $j \to \infty$ we see that $|f_i(x) - f(x)| \leq \varepsilon$ for all $i \geq n(\varepsilon)$ and all x in $[a, b]$. Thus f_i converges to f uniformly on $[a, b]$. By a well-known theorem, it follows that f is continuous on $[a, b]$. Hence $f_i \to f$ in C. This shows that the space (C, ϱ) is complete.

Next consider the same set C, but take for metric the function

$$\sigma(f, g) = \int_a^b |f(x) - g(x)| \, dx.$$

Again it is easy to verify that all the axioms are satisfied. To see that this metric is not topologically equivalent to ϱ, take $f_n(x) = \max(1 - n(x - a), 0)$ and let f be the zero function. Then $\sigma(f_n, f) = 1/2n$ for $n > 1/(b - a)$, but $\varrho(f_n, f) = 1$. Thus $f_n \to f$ in (C, σ) but not in (C, ϱ), hence these spaces are not homeomorphic.

To see that (C, σ) is not complete, take $[a, b] = [0, 1]$, and let

$$f_n(x) = \begin{cases} \min(1, 1/2 - n(x - 1/2)) & \text{on} \quad [0, 1/2] \\ \max(0, 1/2 - n(x - 1/2)) & \text{on} \quad [1/2, 1]. \end{cases}$$

From the graph of this function it is clear that

$$\sigma(f_n, f_m) = (1/4)|1/n - 1/m|.$$

Hence f_n is a Cauchy sequence. Suppose $\sigma(f_n, f) \to 0$ as $n \to \infty$ for some f in C. Then

$$\sigma(f_n, f) \geq \int_0^{1/2 - 1/2n} |1 - f(x)|\, dx + \int_{1/2 + 1/2n}^1 |f(x)|\, dx.$$

Letting $n \to \infty$ it follows that

$$\int_0^{1/2} |1 - f(x)|\, dx = \int_{1/2}^1 |f(x)|\, dx = 0.$$

Since f is continuous, we must have $f(x) = 1$ on $[0, 1/2]$ and $f(x) = 0$ on $[1/2, 1]$, which is impossible.

Next consider the set $R[a, b]$ of Riemann-integrable functions on $[a, b]$, with the same metric σ. Here we encounter a difficulty: the fourth axiom is not satisfied. A set X with a distance function that satisfies only the first three axioms is called a *pseudo metric space*. Such a space can always be made into a metric space by identifying points x and y whenever $\varrho(x, y) = 0$. It is easy to verify that this defines an equivalence relation in X, and that the value of $\varrho(x, y)$ depends only on the equivalence classes to which x and y belong. If we take these classes as the elements of a set $\tilde{X}$, then $(\tilde{X}, \varrho)$ is a metric space. In particular, if we identify any two elements of $R[a, b]$ that differ by a *null function*, that is, a function f such that $\int_a^b |f(x)|\, dx = 0$, we obtain a metric space $(\tilde{R}, \sigma)$, called the *space of R-integrable functions* on $[a, b]$. Let $\tilde{f}$ denote the equivalence class to which f belongs. The mapping $f \to \tilde{f}$ of (C, σ) into $(\tilde{R}, \sigma)$ is distance-preserving. Thus $(\tilde{R}, \sigma)$ contains a subset *isometric* to (C, σ). It is a proper subset, for if f is the indicator function of $[a, (a + b)/2]$, then $\tilde{f}$ belongs to $\tilde{R}$ but no member of $\tilde{f}$ is continuous.

For any positive integer M, let

$$E_M = \{\tilde{f} : f \in R[a, b] \quad \text{and} \quad |f| \leq M\}.$$

Since every integrable function is bounded, $\tilde{R} = \bigcup_{M=1}^{\infty} E_M$. For any $\tilde{f}_0 \in E_M$, with $|f_0| \leq M$, let $g = f_0 + (2M + 1)\chi_I$, where χ_I is the indicator function of an interval I of length ε contained in $[a, b]$. Then $\sigma(\tilde{f}_0, \tilde{g}) = (2M + 1)\varepsilon$. If $|f| \leq M$ then $|g - f| \geq 1$ on I, and so $\sigma(\tilde{f}, \tilde{g}) \geq \varepsilon$. Hence no element of the ε-neighborhood of $\tilde{g}$ belongs to E_M. Since $\tilde{g}$ can be taken arbitrarily close to $\tilde{f}_0$, this shows that E_M is nowhere dense in $\tilde{R}$. Hence $\tilde{R}$ is of first category in itself. It follows that $\tilde{R}$ is not a complete space. Moreover, no remetrization can make it complete, since category is a topological property.

As a last example, consider the class S of sets of finite measure in any measure space, and define

$$\varrho(E, F) = m(E \triangle F).$$

The first two axioms for a metric are obviously satisfied, and the triangle inequality holds, since

$$\varrho(E, H) = m(E \triangle H) = m((E \triangle F) \triangle (F \triangle H))$$
$$\leqq m(E \triangle F) + m(F \triangle H) = \varrho(E, F) + \varrho(F, H).$$

The fourth axiom holds when we identify sets that differ by a nullset. Thereby we obtain a metric space $(\tilde{S}, \varrho)$. To show that this space is complete, let $\tilde{E}_n$ be any Cauchy sequence in $(\tilde{S}, \varrho)$. Then for each positive integer i there is an index n_i such that $\varrho(E_n, E_m) < 1/2^i$ for all $n, m \geqq n_i$, and we may assume that $n_i < n_{i+1}$. Putting $F_i = E_{n_i}$ we have $\varrho(F_i, F_j) < 1/2^i$ for all $j > i$. Define

$$H_i = \bigcap_{j=i}^{\infty} F_j \quad \text{and} \quad E = \bigcup_{i=1}^{\infty} H_i.$$

All of these sets belong to S. E is the set of points that belong to all but a finite number of the sets $F_1, F_2, \ldots$. It is easy to verify that both $E \triangle H_i$ and $H_i \triangle F_i$, and therefore $E \triangle F_i$, are contained in the set

$$(F_i \triangle F_{i+1}) \cup (F_{i+1} \triangle F_{i+2}) \cup (F_{i+2} \triangle F_{i+3}) \cup \cdots.$$

Consequently,

$$m(E \triangle F_i) \leqq \sum_{j=i}^{\infty} m(F_j \triangle F_{j+1}) < \sum_{j=i}^{\infty} 1/2^j = 1/2^{i-1}.$$

For any $n \geqq n_i$, we have

$$m(E \triangle E_n) = m((E \triangle F_i) \triangle (E_{n_i} \triangle E_n))$$
$$\leqq m(E \triangle F_i) + m(E_{n_i} \triangle E_n) < 1/2^{i-1} + 1/2^i.$$

It follows that $\tilde{E}_n$ converges to $\tilde{E}$ in $(\tilde{S}, \varrho)$.

We remark that when m is taken to be 2-dimensional Lebesgue measure in the plane, the space $(\tilde{S}, \varrho)$ contains a subset isometric to $(\tilde{R}, \sigma)$. For any real-valued function f on $[a, b]$, let $\phi(f)$ denote its *ordinate set*, that is, the set

$$\phi(f) = \{(x, y) : a \leqq x \leqq b, \ 0 \leqq y \leqq f(x) \ \text{or} \ f(x) \leqq y \leqq 0\}.$$

It is not hard to see that if f and g belong to $R[a, b]$, then

$$\int_a^b |f - g| = m(\phi(f) \triangle \phi(g)).$$

Hence ϕ takes equivalent functions into equivalent sets and defines an isometric embedding of $(\tilde{R}, \sigma)$ in $(\tilde{S}, \varrho)$. It is possible to identify the closure of $\phi(\tilde{R})$ in $(\tilde{S}, \varrho)$ with the space L^1 of Lebesgue integrable functions on $[a, b]$. This is hardly the easiest way to introduce Lebesgue integration, but it provides one motivation for enlarging the class of integrable functions. Since $(\tilde{R}, \sigma)$ is of first category in itself, it is of first category in any space that contains it topologically. In particular, $\tilde{R}$ is of first category in the space of Lebesgue integrable functions [23].

11. Nowhere Differentiable Functions

Many examples of nowhere differentiable continuous functions are known, the first having been constructed by Weierstrass. One of the simplest existence proofs is due to Banach (1931) [18, p. 327]. It is based on the category method. Banach showed that, in the sense of category, almost all continuous functions are nowhere differentiable; in fact, it is exceptional for a continuous function to have a finite one-sided derivative, or even to have bounded difference quotients on either side, anywhere in an interval.

In the space C of continuous functions on $[0, 1]$, with the uniform metric, let E_n denote the set of functions f such that for some x in $[0, 1 - 1/n]$ the inequality $|f(x + h) - f(x)| \leq nh$ holds for all $0 < h < 1 - x$. To see that E_n is closed, consider any f in the closure of E_n, and let $\{f_k\}$ be a sequence in E_n that converges to f. There is a corresponding sequence of numbers x_k such that, for each k,

$$(1) \qquad 0 \leq x_k \leq 1 - 1/n$$

and

$$(2) \qquad |f_k(x_k + h) - f_k(x_k)| \leq nh \quad \text{for all} \quad 0 < h < 1 - x_k.$$

We may assume also that

$$(3) \qquad x_k \to x, \quad \text{for some} \quad 0 \leq x \leq 1 - 1/n,$$

since this condition will be satisfied if we replace $\{f_k\}$ by a suitably chosen subsequence. If $0 < h < 1 - x$, the inequality $0 < h < 1 - x_k$ holds for all sufficiently large k, and then

$$|f(x + h) - f(x)| \leq |f(x + h) - f(x_k + h)| + |f(x_k + h) - f_k(x_k + h)|$$
$$+ |f_k(x_k + h) - f_k(x_k)| + |f_k(x_k) - f(x_k)| + |f(x_k) - f(x)|$$
$$\leq |f(x + h) - f(x_k + h)| + \varrho(f, f_k) + nh + \varrho(f_k, f) + |f(x_k) - f(x)|.$$

Letting $k \to \infty$, and using the fact that f is continuous at x and $x + h$, it follows that

$$|f(x + h) - f(x)| \leq nh \quad \text{for all} \quad 0 < h < 1 - x.$$

Therefore f belongs to E_n.

$\sigma(x_n, x_m) < 1/2^i$ for all $n, m \geq N$. For all $n, m \geq N$, we have

$$1 > 2^i \sigma(x_n, x_m) \geq \min(1, |f_i(x_n) - f_i(x_m)|),$$

and therefore

$$|f_i(x_n) - f_i(x_m)| < 1.$$

Hence the sequence $\{f_i(x_n)\}$ is bounded, for each i. Since $\varrho(x, y) \leq \sigma(x, y)$, the sequence $\{x_n\}$ is also Cauchy relative to ϱ. Therefore, by hypothesis, the sequence $\{x_n\}$ is convergent. □

Proof of Theorem 12.1. Let X be a non-empty G_δ subset of a complete metric space (Y, ϱ), say $X = \bigcap G_i$, G_i open in Y. Put $F_i = Y - G_i$ and let

$$d(x, F_i) = \inf\{\varrho(x, y) : y \in F_i\}.$$

We may assume that each of the sets F_i is non-empty. Then $d(x, F_i)$ is a real-valued continuous function on Y, positive on X. The functions $f_i(x) = 1/d(x, F_i)$ $(i = 1, 2, ...)$ satisfy the hypotheses of Lemma 12.2. For suppose that $\{x_n\}$ is a Cauchy sequence of points of X, and that for each i the sequence $\{f_i(x_n)\}$ is bounded. Then x_n converges in Y to some point y, since Y is complete. The point y cannot belong to F_i, because then $f_i(x_n) \geq 1/\varrho(x_n, y)$ would be unbounded. Hence $y \in G_i$ for every i; that is, $y \in X$. The sequence $\{x_n\}$ is therefore convergent in the subspace X. Consequently, by Lemma 12.2, X can be remetrized so as to be complete. □

The converse of Alexandroff's theorem is also true, in the following form.

Theorem 12.3. *If a subset X of a metric space (Z, ϱ) is homeomorphic to a complete metric space (Y, σ), then X is a G_δ subset of Z.*

Proof. Let f be a homeomorphism of X onto Y. For each $x \in X$, and each n, there is a positive number $\delta(x, n)$ such that $\sigma(f(x), f(x')) < 1/n$ whenever $\varrho(x, x') < \delta(x, n)$ and $x' \in X$. We may assume that $\delta(x, n) < 1/n$. Let G_n be the union of the balls in Z with center x and radius $\delta(x, n)/2$, the union being taken as x ranges over X. Then G_n is open in Z. Let $z \in \bigcap G_n$. For each n there is a point $x_n \in X$ such that $\varrho(z, x_n) < \delta(x_n, n)/2$. Since $\delta(x_n, n) < 1/n$ it follows that $x_n \to z$. Also, for any $m > n$, we have

$$\varrho(x_n, x_m) \leq \varrho(z, x_n) + \varrho(z, x_m) < \delta(x_n, n)/2 + \delta(x_m, m)/2 \leq \delta(x_n, n) \text{ or } \delta(x_m, m).$$

Therefore $\sigma(f(x_n), f(x_m)) < 1/n$ for all $m > n$. The sequence $y_n = f(x_n)$ is therefore Cauchy in (Y, σ). Hence it is convergent, say $y_n \to y$. Put $x = f^{-1}(y)$. Then $x \in X$ and $x_n \to x$, because f^{-1} is continuous. Since x_n has already been shown to converge to z, it follows that $z = x$, and therefore $z \in X$. This proves that $\bigcap G_n \subset X$. The opposite inclusion is obvious from the definition of G_n. Hence X is a G_δ subset of Z. □

13. Transforming Linear Sets into Nullsets

Let F be a nowhere dense closed subset of $I = [0, 1]$. It is an almost trivial observation that there exists a homeomorphism h of I onto itself such that $h(F)$ is a nullset. In fact, letting $G = I - F$, it suffices to take

$$h(x) = m([0, x] \cap G)/m(G).$$

This is a strictly increasing continuous map of I onto itself. The intervals that compose G are mapped onto a sequence of intervals of total length 1. Hence $h(F)$ is a nullset.

The generalization of this result to a set of first category cannot be proved in the same way. Nevertheless, the conclusion still holds, as we shall show by a category argument.

Let H denote the set of all automorphisms of I (that is, homeomorphisms of I onto itself) that leave the endpoints fixed, metrized by the distance function

$$\varrho(g, h) = \max |g(x) - h(x)|.$$

Evidently, (H, ϱ) is a subspace of the space $C = C[0, 1]$ of continuous functions on I, and also of the subspace C_1 of continuous mappings of I into R that leave 0 and 1 fixed. The space (C_1, ϱ) is complete, since C_1 is a closed subset of C. However, the space (H, ϱ) is not complete. This may be seen by considering the sequence $\{f_n\}$, where f_n is the piece-wise linear function whose graph consists of the line segments joining the point $(1/2, 1 - 1/n)$ to $(0, 0)$ and to $(1, 1)$.

Let H_n be the set of all f in C such that $f(x) \neq f(y)$ for all x and y in I with $|x - y| \geq 1/n$. if f belongs to H_n, then the number

$$\delta = \min \{|f(x) - f(y)| : |x - y| \geq 1/n\}$$

is positive. If $\varrho(f, g) < \delta/2$, then

$$|g(x) - g(y)| \geq |f(x) - f(y)| - 2\varrho(f, g) \geq \delta - 2\varrho(f, g) > 0$$

whenever $|x - y| \geq 1/n$, and so g belongs to H_n. This shows that H_n is an open subset of C. Evidently

$$H = C_1 \cap \bigcap H_n.$$

Hence H is a G_δ subset of C, and therefore topologically complete, by Theorem 12.1. (It can be shown that H is complete with respect to the metric

$$\sigma(g, h) = \varrho(g, h) + \varrho(g^{-1}, h^{-1}),$$

and that this metric is topologically equivalent to ϱ, but we shall not make use of this fact.)

Theorem 13.1. *For any set A of first category in $I = [0, 1]$ there exists an $h \in H$ such that $h(A)$ is a nullset; indeed, such automorphisms constitute a residual set in H.*

Proof. Let $A = \bigcup A_n$, A_n nowhere dense. Let

$$E_{n,k} = \{h \in H : m(h(\overline{A}_n)) < 1/k\}\ .$$

For any $h \in E_{n,k}$, the bounded closed set $h(\overline{A}_n)$ can be enclosed in an open set G in R such that $m(G) < 1/k$. There exists a number $\delta > 0$ such that G contains the δ-neighborhood of each point of $h(\overline{A}_n)$. If $\varrho(g, h) < \delta$, then $g(\overline{A}_n) \subset G$, and therefore g belongs to $E_{n,k}$. This shows that $E_{n,k}$ is an open subset of H, for all n and k.

For any $g \in H$ and $\varepsilon > 0$, divide I into a finite number of closed sub-intervals $I_1, \ldots, I_N$ of length less than ε. In the interior I_i^0 of I_i, choose a closed interval $J_i \subset I_i^0 - g(\overline{A}_n)$ $(i = 1, \ldots, N)$. Let h_i be a piece-wise linear homeomorphism of I_i onto itself that leaves the endpoints fixed and maps J_i onto an interval of length greater than $|I_i| - 1/kN$. (Three line segments suffice to define the graph of such a function h_i.) Together, these h_i define a mapping $h \in H$ such that $m(h \circ g(\overline{A}_n)) < 1/k$. Therefore $h \circ g$ belongs to $E_{n,k}$. Since $\varrho(h \circ g, g) < \varepsilon$, it follows that $E_{n,k}$ is dense in H. Consequently, the set

$$E = \bigcap_{n,k} E_{n,k}$$

is a residual set in H. If $h \in E$, then $h(\overline{A}_n)$ is a nullset for every n. Since $h(A) \subset \bigcup h(\overline{A}_n)$, it follows that $h(A)$ is a nullset. $\square$

The following theorem gives a sufficient condition for the opposite conclusion.

Theorem 13.2. *For any uncountable closed set A contained in $I = [0, 1]$, there exists an $h \in H$ such that $h(A)$ has positive measure.*

Proof. By Lemma 5.1, there exists a closed set $F \subset A$ and a continuous map f of F onto $[0, 1]$. For each $x \in I$, define

$$h(x) = x/2 + m(f([0, x] \cap F))/2\ .$$

Then h is a strictly increasing continuous map of $[0, 1]$ onto itself. On each of the open intervals that compose $(0, 1) - F$, we have $h'(x) = 1/2$. Hence $m(h(I - F)) = 1/2$, and therefore $m(h(A)) \geq m(h(F)) = 1/2$. $\square$

The preceding theorems have an amusing consequence. Let f be a bounded function on $[0, 1]$, and let D be its set of points of discontinuity. Let h be an arbitrary automorphism of $[0, 1]$. Then the composite function $f \circ h$ is bounded and has the set $h^{-1}(D)$ for its set of points of discontinuity. We know (Theorem 7.1) that D is always an F_σ. If D is uncountable, it contains an uncountable closed set. Then, for some h, the set $h^{-1}(D)$ has positive measure. On the other hand, if D is countable then $h^{-1}(D)$ is countable and has measure zero for every h. If D is of first category, there exists an h such that $h^{-1}(D)$ is a nullset. On the other hand, if D is of second category then one of the closed sets of which it is the union must contain an interval. In this case, $h^{-1}(D)$ also contains an interval, and is never a nullset. Recalling Theorem 7.5, that a bounded function is R-integrable if and only if it is continuous almost everywhere, we have the following

Theorem 13.3. *Let f be a bounded function on $[0, 1]$, and let D be its set of points of discontinuity. Let h be an arbitrary homeomorphism of $[0, 1]$ onto itself. Then the composite function $f \circ h$ is Riemann integrable*

(a) *for all h if and only if D is countable,*

(b) *for some h if and only if D is of first category,*

(c) *for the identity mapping h if and only if D is a nullset.*

In this theorem each of the σ-ideals we have been considering answers a question concerning the effect of a strictly monotone substitution on the Riemann integrability of a function!

Another consequence is the following characterization of sets of first category, in which the notion of a nowhere dense set does not appear.

Theorem 13.4. *A linear set A is of first category if and only if there exists a homeomorphism h of the line onto itself such that $h(A)$ is contained in an F_σ nullset.*

This theorem characterizes sets of first category as those that are topologically equivalent to a special kind of nullset.

Proof. Any set A of first category is contained in an F_σ set B of first category. Divide the line into non-overlapping intervals I_i of unit length. Let h_i be an automorphism of I_i that leaves the endpoints fixed and maps $B \cap I_i$ onto a nullset. The mappings h_i define an automorphism h of the line such that $h(B)$ is a nullset. Therefore $h(A)$ is contained in the F_σ nullset $h(B)$.

Conversely, let A be any subset of an F_σ nullset. Then $A \subset \bigcup F_n$, where F_n is a closed nullset. Therefore each of the sets F_n is nowhere dense. Consequently A, and its image under any automorphism of the line, is of first category. $\square$

14. Fubini's Theorem

Linear Lebesgue measure is defined by covering sequences of intervals, and plane measure by covering sequences of rectangles. We shall now consider how these measures are related to each other. It is clear what kind of answer we should expect. In elementary calculus we learn to compute the area between the graphs of two functions $f \leq g$ by the formula

$$\int_a^b |f(x) - g(x)| \, dx \, .$$

Thus the area is computed "by slicing." The generalization of this formula, which expresses the measure of any plane measurable set A as the integral of the linear measure of its sections perpendicular to an axis, is called Fubini's theorem. We shall not formulate the theorem in full generality, but confine attention to the case in which A is a nullset. Then the theorem asserts that almost all vertical (or horizontal) sections of A have measure zero.

Let X and Y be two non-empty sets. For any $A \subset X$ and $B \subset Y$, the *product set* $A \times B$ is defined to be the set of all ordered pairs (x, y), where $x \in A$ and $y \in B$. For instance, in coordinate geometry the plane is represented as the product of two lines. If $E \subset X \times Y$ and $x \in X$, the set

$$E_x = \{y : (x, y) \in E\}$$

is called the *x-section* of E. Note that E_x is a subset of Y, not of $X \times Y$. The operation of sectioning commutes with union, intersection, and complement. That is,

$$(E \cup F)_x = E_x \cup F_x, \quad (E \cap F)_x = E_x \cap F_x, (E')_x = (E_x)' \, .$$

Thus the mapping $E \to E_x$, for any fixed $x \in X$, is a homomorphism of the Boolean algebra of subsets of $X \times Y$ onto the algebra of subsets of Y. It is even a homomorphism with respect to arbitrary union and intersection; that is,

$$\left(\bigcup E_i \right)_x = \bigcup \left[(E_i)_x \right] \quad \text{and} \quad \left(\bigcap E_i \right)_x = \bigcap \left[(E_i)_x \right] \, .$$

A set A is said to be *covered infinitely many times* by the sequence $\{A_n\}$ if each point of A belongs to infinitely many terms of the sequence.

It is sometimes convenient to use the following characterization of nullsets.

Lemma 14.1. *A set A has Lebesgue measure zero if and only if it can be covered infinitely many times by a sequence of intervals I_n such that the series $\sum |I_n|$ is convergent.*

Proof. If A is a nullset, then it can be covered by a sequence of intervals the sum of whose measures is less than 1/2, by another with sum less than 1/4, by another with sum less than 1/8, and so on. Together these constitute a sequence $\{I_n\}$ that covers A infinitely many times, and $\sum |I_n| < 1$.

Conversely, if A is covered infinitely many times by a sequence $\{I_n\}$, then it is covered by the subsequence starting with the k-th term. If $\sum |I_n|$ is convergent, the sum $\sum_k^\infty |I_n|$ can be made arbitrarily small by suitable choice of k. Hence A is a nullset. $\square$

Theorem 14.2 (Fubini). *If E is a plane set of measure zero, then E_x is a linear nullset for all x except a set A of linear measure zero.*

Proof. For any $\varepsilon > 0$ let $I_i \times J_i$ be a sequence of rectangles, with I_i and J_i half open on the left, such that

(1) the sequence $I_i \times J_i$ covers E infinitely many times,

and

(2) $$\sum_i |I_i| \, |J_i| \leq \varepsilon .$$

By further subdividing each interval I_i we can insure also that

(3) for each $i > 1$, I_i is contained in a single interval of the subdivision of the line determined by the endpoints of the intervals $I_1, I_2, \ldots, I_{i-1}$.

Define $\phi_0(x) = 0$ and

$$\phi_n(x) = \sum_{x \in I_i, i \leq n} |J_i| \quad (n = 1, 2, \ldots) .$$

Then ϕ_i is a step function, $\phi_{i-1} \leq \phi_i$, and

$$\phi_i(x) - \phi_{i-1}(x) = \begin{cases} |J_i| & \text{for} \quad x \in I_i \\ 0 & \text{elsewhere} . \end{cases}$$

Hence, by (2),

(4) $$\int \phi_n \, dx = \sum_1^n \int (\phi_i - \phi_{i-1}) \, dx = \sum_1^n |I_i| \, |J_i| \leq \varepsilon .$$

Let

$$A_i = \{x : \phi_i(x) \geq 1 > \phi_{i-1}(x)\} \quad (i = 1, 2, \ldots) .$$

Then A_i is either empty or equal to I_i, and the intervals A_i are disjoint. Since $\phi_n(x) \geqq 1$ on A_i for $n \geqq i$, we have

$$\sum_1^n |A_i| \leqq \int \phi_n \, dx \quad \text{for each } n,$$

and therefore, by (4),

$$(5) \qquad\qquad \sum_1^\infty |A_i| \leqq \varepsilon.$$

Let $A = \{x : E_x \text{ is not a nullset}\}$. For each $x \in A$ we have $(x, y) \in E$ for some y, and therefore $(x, y) \in I_i \times J_i$ for infinitely many i. Let $\{i_k\}$ be the sequence of indices such that $x \in I_{i_k}$. If $y \in E_x$, then $y \in J_{i_k}$ for infinitely many k, by (1). Thus, the sequence $\{J_{i_k}\}$ covers E_x infinitely many times. Since E_x is not a nullset, the series $\sum |J_{i_k}|$ must diverge. Hence, for each $x \in A$, we have $\lim_{n \to \infty} \phi_n(x) = \infty$, and therefore $x \in A_i$ for some i. This shows that the sequence of intervals $A_1, A_2, \ldots$ covers the set A. From (5), it follows that A is a linear nullset. $\square$

Theorem 14.3. *If E is a plane measurable set, then E_x is linearly measurable for all x except a set of linear measure zero.*

Proof. By Theorem 3.15, E can be represented as the union of an F_σ set A and a nullset N. We have $E_x = A_x \cup N_x$ for all x. Any section of a closed set is closed, hence A_x is an F_σ for every x. By Fubini's theorem, N_x is a nullset for almost all x. Since E_x is measurable for any such x, the conclusion follows. $\square$

The converse of Fubini's theorem is true in the sense that if almost all sections of a plane measurable set E are nullsets, then E is a nullset. We shall not prove this. (The usual proof depends on properties of the Lebesgue integral, which we have not developed here.) We remark only that the conclusion does not follow unless E is assumed to be measurable. This is shown by the following theorem, due to Sierpinski.

Theorem 14.4. *There exists a plane set E such that* (a) *E meets every closed set of positive plane measure, and* (b) *no three points of E are collinear.*

Such a set E cannot be measurable. For if E were measurable, then (a) and Theorem 3.18 would imply that its complement is a nullset, in which case (b) would contradict Fubini's theorem. Hence E is not measurable, and therefore not a nullset, despite the fact that each of its sections has at most two points.

For Sierpinski's proof of the above theorem, see Fund. Math. Vol. 1, p. 112. We shall give here only a simplified version, assuming the continuum hypothesis.

Let the class of closed sets of positive plane measure be well ordered in such a way that each member has only countably many predecessors.

This is possible, assuming the continuum hypothesis, since the class of closed sets of positive measure has power c. Choose a point p_1 of the first set F_1, then a point p_2 of the next set F_2, with $p_2 \neq p_1$. Then choose p_3 in F_3 not collinear with p_1 and p_2. Assuming points have been selected from each of the sets preceding F_α, choose p_α in F_α in such a way that p_α is not collinear with any two of the points already chosen. Only countably many points have indices less than α, so only countably many lines have to be avoided in choosing p_α. The union of these lines has plane measure zero, and F_α has positive plane measure. Hence such a point p_α can always be found. The totality of points p_α so chosen defines a set E having properties (a) and (b).

15. The Kuratowski-Ulam Theorem

Fubini's theorem has a category analogue. In its general formulation, this theorem was proved in 1932 by Kuratowski and Ulam [18, p. 222].

Theorem 15.1 (Kuratowski-Ulam). *If E is a plane set of first category, then E_x is a linear set of first category for all x except a set of first category. If E is a nowhere dense subset of the plane $X \times Y$, then E_x is a nowhere dense subset of Y for all x except a set of first category in X.*

Proof. The two statement are essentially equivalent. For if $E = \bigcup E_i$, then $E_x = \bigcup_i (E_i)_x$. Hence the first statement follows from the second. If E is nowhere dense, so is $\bar{E}$, and E_x is nowhere dense whenever $(\bar{E})_x$ is of first category. Hence the second statement follows from the first. It is therefore sufficient to prove the second statement for any nowhere dense closed set E.

Let $\{V_n\}$ be a countable base for Y, and put $G = (X \times Y) - E$. Then G is a dense open subset of the plane. For each positive integer n, let G_n be the projection of $G \cap (X \times V_n)$ in X, that is,

$$G_n = \{x : (x, y) \in G \text{ for some } y \in V_n\}.$$

Let $x \in G_n$ and $y \in V_n$ be such that $(x, y) \in G$. Since G is open, there exist open intervals U and V such that $x \in U$, $y \in V \subset V_n$, and $U \times V \subset G$. It follows that $U \subset G_n$. Hence G_n is an open subset of X. For any non-empty open set U, the set $G \cap (U \times V_n)$ is non-empty, since G is dense in the plane. Hence G_n contains points of U. Therefore G_n is a dense open subset of X, for each n. Consequently, the set $\bigcap G_n$ is the complement of a set of first category in X. For any $x \in \bigcap G_n$, the section G_x contains points of V_n for every n. Hence G_x is a dense open subset of Y, and therefore $E_x = Y - G_x$ is nowhere dense. This shows that for all x except a set of first category, E_x is nowhere dense. $\square$

The proofs of this and of the next three theorems apply to the Cartesian product $X \times Y$ of any two topological spaces, provided only that Y has a countable base. In fact, it is sufficient to assume that there is a sequence of non-empty open sets in Y such that every non-empty open set contains a member of the sequence.

Theorem 15.2. *If E is a subset of $X \times Y$ with the property of Baire, then E_x has the property of Baire for all x except a set of first category in X.*

Proof. Let $E = G \triangle P$, where G is open and P is of first category. Then $E_x = G_x \triangle P_x$, for all x. Every section of an open set is open, hence E_x has the property of Baire whenever P_x is of first category. By Theorem 15.1, this is the case for all x except a set of first category. $\square$

Theorem 15.3. *A product set $A \times B$ is of first category in $X \times Y$ if and only if at least one of the sets A or B is of first category.*

Proof. If G is a dense open subset of X, then $G \times Y$ is a dense open subset of $X \times Y$. Hence $A \times B$ is nowhere dense in $X \times Y$ whenever A is nowhere dense in X. Since $(\bigcup A_i) \times B = \bigcup (A_i \times B)$, it follows that $A \times B$ is of first category whenever A is of first category. Similar reasoning applies to B.

Conversely, if $A \times B$ is of first category and A is not, then by Theorem 15.1 there exists a point x in A such that $(A \times B)_x$ is of first category. Since $(A \times B)_x = B$ for all x in A, it follows that B is of first category. $\square$

The following theorem is a partial converse of Theorem 15.1.

Theorem 15.4. *If E is a subset of $X \times Y$ that has the property of Baire, and if E_x is of first category for all x except a set of first category, then E is of first category.*

Proof. Suppose the contrary. Then $E = G \triangle P$, where P is of first category and G is an open set of second category. There exist open sets U and V such that $U \times V \subset G$ and $U \times V$ is of second category. (This is clear in the case of the plane. In general it follows from the Banach category theorem, which will be discussed in the next chapter.) By Theorem 15.3, both U and V are of second category. For all x in U, $E_x \supset V - P_x$. By Theorem 15.1, P_x is of first category for all x except a set of first category. Therefore E_x is of second category for all x in U except a set of first category. This implies that E_x is of second category for all x in a set of second category, contrary to hypothesis. $\square$

That Theorem 15.4 is not true without the first hypothesis is shown by the following analogue of Theorem 14.4.

Theorem 15.5. *There exists a plane set E of second category such that no three points of E are collinear.*

Proof. The class of plane G_δ sets of second category has power c. Let $\{E_\alpha : \alpha < \omega_c\}$ be a well ordering of this class, where ω_c is the first ordinal preceded by c ordinals. Suppose points p_β, with no three collinear and with $p_\beta \in F_\beta$, have been chosen for all $\beta < \alpha$. Since the set of all lines joining pairs of points p_β with $\beta < \alpha$ has power less than c, we can find a direction not parallel to any of these lines. By Theorem 15.4, some line in this direction meets E_α in a set of second category, and therefore, by Lemma 5.1, in a set of power c. We can therefore choose p_α in E_α in such a way that p_α is not collinear with any two points p_β with $\beta < \alpha$. The set of all points p_α so chosen contains no three collinear points. It is of second category because its complement contains no G_δ set of second category. ▯

The analogy between the statements of the Fubini and Kuratowski-Ulam theorems poses an interesting question: Can either statement be reduced to the other? We shall show that, in a sense, the Kuratowski-Ulam theorem can be reduced to Fubini's. The reduction is limited to the case of plane sets, and it does not lead to any simplification. The interest of the question lies in the technical problem it poses: to find a transformation of the plane that will reduce any given instance of the Kuratowski-Ulam theorem to a case of Fubini's. No similar reduction of Fubini's theorem to that of Kuratowski-Ulam appears possible.

In Chapter 13 it was shown that any linear set of first category can be transformed into a nullset by an automorphism of the line. Similarly, it can be shown that any set of first category in r-space can be transformed into one of measure zero by an automorphism of the space [27]. This was first proved in 1919 (for subsets of the square) by L. E. J. Brouwer [6]. However, this result is inadequate for our present purpose, because such a transformation need not take sections into sections. What is needed is a stronger version of Brouwer's theorem, asserting that the automorphism can be taken to be a *product transformation* $f \times g$, that is, one of the form $x' = f(x)$, $y' = g(y)$. We shall establish the existence of such an automorphism by a category argument similar to the one we gave in the 1-dimensional case.

Let m_2 and m denote 2-dimensional and linear Lebesgue measure, respectively.

Theorem 15.6. *For any plane set E of first category contained in the unit square, there exists a product homeomorphism h of the unit square onto itself such that $m_2(h(E)) = 0$.*

Proof. As in Chapter 13, let (H, ϱ) denote the space of automorphisms of the unit interval that leave the endpoints fixed. Let H^2 denote the set of all automorphisms of the unit square of the form $f \times g$, where f and g belong to H. H^2 may be identified with the Cartesian product $H \times H$. If $h_1 = f_1 \times g_1$ and $h_2 = f_2 \times g_2$, define

$$\sigma(h_1, h_2) = \varrho(f_1, f_2) + \varrho(g_1, g_2).$$

It is easy to verify that (H^2, σ) is a metric space, and that it is topologically complete. (Any remetrization of H determines a corresponding remetrization of H^2.)

Let F be a nowhere dense closed subset of the square. For each positive integer k, define

$$E_k = \{h \in H^2 : m_2(h(F)) < 1/k\} .$$

By the same reasoning as in the proof of Theorem 13.1, we see that E_k is an open subset of H^2. To show that it is dense, let ε be any positive number, choose $n > 2/\varepsilon$, and divide the unit interval into n equal closed subintervals

$$I_i = [(i-1)/n, i/n] \qquad (i = 1, 2, ..., n) .$$

Let $F_{ij} = F \cap (I_i \times I_j)$, and let T_{ij} denote the translation

$$x' = x - (i-1)/n , \qquad y' = y - (j-1)/n .$$

Then the finite union $\bigcup_{i,j} T_{ij}(F_{ij})$ is a nowhere dense subset of $I_1 \times I_1$. Choose closed intervals J and K interior to I_1 such that

$$J \times K \subset (I_1 \times I_1) - \bigcup T_{ij}(F_{ij}) .$$

Then

$$[(i-1) + J] \times [(j-1) + K] \subset (I_i \times I_j) - F .$$

In each of the squares $I_i \times I_j$, F has a similarly situated hole. Here is an illustration of the case $n = 3$.

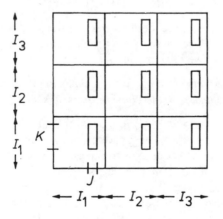

Let f_1 be a piece-wise linear automorphism of I_1, leaving the endpoints fixed, such that

$$m(f_1(J)) > \sqrt{1 - 1/k} \, m(I_1) .$$

(Three linear pieces suffice.) Similarly, let g_1 be a piece-wise linear automorphism of I_1, leaving the endpoints fixed, such that

$$m(g_1(K)) > \sqrt{1 - 1/k}\, m(I_1)\,.$$

Define

$$f_i(x) = \frac{i-1}{n} + f_1\left(x - \frac{i-1}{n}\right) \quad \text{for} \quad x \in I_i\,,$$

and

$$g_j(y) = \frac{j-1}{n} + g_1\left(y - \frac{j-1}{n}\right) \quad \text{for} \quad y \in I_j\,.$$

The transformations f_i $(i = 1, ..., n)$ together define a piecewise linear transformation $f \in H$. Similarly, the transformations g_j define a transformation $g \in H$. The product transformation $f \times g$ maps each square $I_i \times I_j$ onto itself, hence its σ-distance from the identity is less than ε. Moreover, $m_2(h(F)) < 1/k$, since the part of F contained in $I_i \times I_j$ is mapped onto a set of measure less than $1/kn^2$. Thus, for any nowhere dense closed set F, the ε-neighborhood of the identity in H^2 contains points of E_k. When this result is applied to the set $\phi(F)$, for any given $\phi \in H^2$, we obtain an element $h \in H^2$ such that $m_2(h \circ \phi(F)) < 1/k$ and $\sigma(h \circ \phi, \phi) < \varepsilon$. This shows that E_k is dense in H^2, in addition to being open.

If E is any set of first category in the unit square, then $E \subset \bigcup F_i$, where F_i is closed and nowhere dense. For any two positive integers i and k let

$$E_{ik} = \{h \in H^2 : m_2(h(F_i)) < 1/k\}\,.$$

We have shown that E_{ik} is a dense open subset of the topologically complete space H^2. Hence there exists an element $h \in \bigcap_{i,k} E_{ik}$. For any such automorphism h, we have $m_2(h(E)) = 0$. $\square$

Theorem 15.7. *For any set E of first category in the plane, there exists a product homeomorphism h of the plane onto itself such that $m_2(h(E)) = 0$.*

Proof. Let $f(x) = \tan \pi(x - 1/2)$ $(0 < x < 1)$. Then f is a homeomorphism of $(0, 1)$ onto the line. The product homeomorphism $g = f \times f$ maps the interior of the unit square onto the plane. Since f' is continuous and positive, both g and g^{-1} map nullsets onto nullsets. If E is of first category in the plane, then $g^{-1}(E)$ is a set of first category in the square. By Theorem 15.6, there exist an automorphism $h \in H^2$ such that $h \circ g^{-1}(E)$ is a nullset. Then $g \circ h \circ g^{-1}$ is a product automorphism of the plane that maps E onto a nullset. $\square$

It is now an easy matter to reduce any instance of the Kuratowski-Ulam theorem to Fubini's theorem. Let E be a nowhere dense closed subset of the plane. Then each section E_x is either nowhere dense or it contains an interval. Let $V_1, V_2, \ldots$ be an enumeration of all open intervals with rational endpoints. Put $F_i = \{x : E_x \supset V_i\}$. F_i is a closed subset of the line, since each horizontal section $E^y = \{x : (x, y) \in E\}$ of E is closed, and $F_i = \bigcap_{y \in V_i} E^y$. The F_σ set $A = \bigcup F_i$ is the set of all points x for which E_x is not of first category.

Let $h = f \times g$ be a product homeomorphism of the plane onto itself such that $m_2(h(E)) = 0$ (Theorem 15.7). For each x in A, the section E_x contains some interval V_i. The section $(h(E))_{f(x)}$ contains the interval $g(V_i)$, and so is not a nullset. Therefore $f(A) \subset B$, where

$$B = \{x : (h(E))_x \text{ is not a nullset}\} .$$

By Fubini's theorem, $m(B) = 0$. Thus $f(A)$ is an F_σ set of measure zero. Hence $f(A)$, and therefore A, is of first category, by Theorem 13.4.

16. The Banach Category Theorem

In a topological space that has a countable base, it is obvious that the union of any family of open sets of first category is of first category. One need only take the union of those members of the base that are contained in at least one member of the given family. The same reasoning shows that the union of any family of open sets of measure zero has measure zero (for any measure defined for all open sets). It is remarkable that the first statement remains valid whether the space has a countable base or not. The second statement, however, needs to be qualified.

Theorem 16.1 (Banach Category Theorem). *In a topological space X, the union of any family of open sets of first category is of first category.*

Proof. Let G be the union of a family $\mathscr{G}$ of non-empty open sets of first category. Let $\mathscr{F} = \{U_\alpha : \alpha \in A\}$ be a maximal family of disjoint non-empty open sets with the property that each is contained in some member of $\mathscr{G}$. Then the closed set $\bar{G} - \bigcup \mathscr{F}$ is nowhere dense. (Otherwise $\mathscr{F}$ would not be maximal.) Each set U_α can be represented as a countable union of nowhere dense sets, say $U_\alpha = \bigcup_{n=1}^\infty N_{\alpha,n}$. Put $N_n = \bigcup_{\alpha \in A} N_{\alpha,n}$. If an open set U meets N_n, then it meets some $N_{\alpha,n}$ and there exists a non-empty open set $V \subset (U \cap U_\alpha) - N_{\alpha,n}$. Hence $V \subset U - N_n$, and so N_n is nowhere dense. Therefore

$$G \subset (\bar{G} - \bigcup\mathscr{F}) \cup \bigcup_{\alpha \in A} U_\alpha = (\bar{G} - \bigcup\mathscr{F}) \cup \bigcup_1^\infty N_n$$

is of first category. □

It follows that any topological space is the union of an open (or closed) Baire subspace and a set of first category. To discuss the analogue of Theorem 16.1 for open sets of measure zero, we need the following lemma, which is due to Montgomery [18, p. 265].

Lemma 16.2 (Montgomery). *Let $\{G_\alpha : \alpha \in A\}$ be a well-ordered family of open subsets of a metric space X, and for each $\alpha \in A$ let F_α be a closed subset of*

$$H_\alpha = G_\alpha - \bigcup_{\beta < \alpha} G_\beta .$$

Then the set $E = \bigcup_{\alpha \in A} F_\alpha$ is an F_σ.

Proof. For each $\alpha \in A$ and each positive integer n, let

$$F_{\alpha,n} = \{x \in F_\alpha : d(x, X - G_\alpha) \geq 1/n\} .$$

Then $F_{\alpha,n}$ is a closed set, and $F_\alpha = \bigcup_{n=1}^\infty F_{\alpha,n}$. If $\alpha \neq \beta$, we have $\varrho(x, y) \geq 1/n$ for every $x \in F_{\alpha,n}$ and $y \in F_{\beta,n}$. Hence any convergent sequence contained in the set $F_n = \bigcup_{\alpha \in A} F_{\alpha,n}$ must, except for a finite number of terms, be contained in a single set $F_{\alpha,n}$. It follows that F_n is closed, and that

$$E = \bigcup_{\alpha \in A} F_\alpha = \bigcup_{n=1}^\infty F_n$$

is an F_σ. ⬚

Following Marczewski and Sikorski [20, 21], to whom the following theorems are due, a cardinal is said to have *measure zero* if every finite measure defined for all subsets of a set of that cardinality vanishes identically if it is zero for points. Evidently any cardinal less than one of measure zero has measure zero. As already mentioned in Chapter 5, it is known that every cardinal less than the first weakly inaccessible cardinal has measure zero, and that (assuming the continuum hypothesis) only exceedingly large cardinals can fail to have measure zero.

A measure μ defined on the class of Borel subsets of a space X is called a *Borel measure*. It is *normalized* if $\mu(X) = 1$, and *nonatomic* if it is zero for points.

Theorem 16.3. *Let μ be a finite Borel measure in a metric space X. If G is the union of a family $\mathscr{G}$ of open sets of measure zero, and if card $\mathscr{G}$ has measure zero, then $\mu(G) = 0$.*

Proof. Let $\{G_\alpha : \alpha \in A\}$ be a well ordering of $\mathscr{G}$, and put $H_\alpha = G_\alpha - \bigcup_{\beta < \alpha} G_\beta$ for each $\alpha \in A$. Each of the sets H_α is the difference of two open sets, therefore an F_σ, say $H_\alpha = \bigcup_{n=1}^\infty F_{\alpha,n}$, where $F_{\alpha,n}$ is closed. For any set $E \subset A$ we have

$$\bigcup_{\alpha \in E} H_\alpha = \bigcup_{\alpha \in E} \bigcup_{n=1}^\infty F_{\alpha,n} = \bigcup_{n=1}^\infty \left[\bigcup_{\alpha \in E} F_{\alpha,n} \right] .$$

By Montgomery's lemma (taking $F_\alpha = \emptyset$ when $\alpha \in A - E$), the set in brackets is an F_σ, for each n, and therefore the union is too. Hence the set function

$$\nu(E) = \mu\left(\bigcup_{\alpha \in E} H_\alpha \right)$$

is defined for all subsets of A. It is evidently a finite measure, and non-atomic. Since card A has measure zero it follows that $\mu(G) = \mu(\bigcup_{\alpha \in A} H_\alpha) = \nu(A) = 0$. ⬚

Theorem 16.4. *If X is a metric space with a base whose cardinal has measure zero, and if μ is a finite Borel measure in X, then the union of any family of open sets of measure zero has measure zero.*

Proof. Let $\mathscr{B}$ be a base whose cardinal has measure zero. For any family $\mathscr{G}$ of open sets of measure zero, let $\mathscr{B}_0$ be the set of all members of

$\mathscr{B}$ that are contained in some member of $\mathscr{G}$. Then $\mu(\bigcup\mathscr{G})=\mu(\bigcup\mathscr{B}_0)=0$, by Theorem 16.3. □

Surprisingly, Theorem 16.4 can fail for measures in nonmetrizable spaces. Kemperman and Maharam [17] have shown that in the Cartesian product X of c copies of the line it is possible to define a normalized measure μ on the σ-algebra generated by the elementary open sets, with the property that X can be covered by a family of open sets of measure zero. The elementary open sets constitute a base of cardinality c, and c has measure zero (assuming the continuum hypothesis) by Ulam's Theorem 5.6.

The following theorem is perhaps the ultimate generalization of Theorem 1.6.

Theorem 16.5. *Let X be a metric space with a base whose cardinal has measure zero. Let μ be a nonatomic Borel measure in X such that*

(i) *every set of infinite measure has a subset with positive finite measure, and*

(ii) *every set of measure zero is contained in a G_δ set of measure zero.*

Then X can be represented as the union of a G_δ set of measure zero and a set of first category.

Proof. By selecting a point from each member of the given base, we obtain a dense set S of at most the same cardinality. For each positive integer n, let F_n be a maximal subset of S with the property that $\varrho(x, y) \geq 1/n$ for any two distinct points of F_n. Put $D = \bigcup_1^\infty F_n$. Then D is dense in X, and since every subset of F_n is closed, every subset of D is an F_σ. Hence μ is defined for all subsets of D. Because μ is zero for points and card D has measure zero, it follows that no subset of D has positive finite measure. Therefore, by (i) and (ii), $\mu(D) = 0$ and D is contained in a G_δ set E with $\mu(E) = 0$. The complement of E is a set of first category. □

The restriction on cardinality is essential. Because if there exists a set X whose cardinal does not have measure zero, it can be metrized by defining $\varrho(x, y) = 1$ for all $x \neq y$. Then all subsets of X are open and a nontrivial finite measure defined for all subsets of X and zero for points would be a Borel measure that satisfies conditions (i) and (ii) but not the conclusion.

It is easy to verify that any finite Borel measure in a metric space satisfies conditions (i) and (ii). (The class of Borel sets that have an F_σ subset and a G_δ superset of equal measure is a σ-algebra that includes all closed sets.) However, these conditions cannot be omitted from Theorem 16.5, as may be seen by considering the Borel measure μ in R defined by putting $\mu(E) = m(E)$ for every Borel set E of first category, and $\mu(E) = \infty$ for every Borel set of second category.

17. The Poincaré Recurrence Theorem

In the course of his studies in celestial mechanics, Poincaré discovered a theorem which is remarkable both for its simplicity and for its far-reaching consequences. It is noteworthy also for having initiated the modern study of measure-preserving transformations, known as ergodic theory. From our point of view, this "recurrence theorem" has a special interest, because in proving it Poincaré anticipated the notions of both measure and category. Publication of his treatise, "Les méthodes nouvelles de la mécanique céleste" [29], antedated slightly the introduction of either notion.

Let X be a bounded open region of r-space, and let T be a homeomorphism of X onto itself that preserves volume; that is, G and $T(G)$ have equal volume, for every open set $G \subset X$. Under iteration of T, each point x generates a sequence $x, Tx, T^2x, \ldots, T^ix, \ldots$ called the *positive semiorbit* of x. A point x of an open set G is said to be *recurrent with respect to G* if T^ix belongs to G for infinitely many positive integers i. In effect, Poincaré proved two theorems, which may be stated together as follows.

Theorem 17.1. *For any open set $G \subset X$, all points of G are recurrent with respect to G except a set of first category and measure zero.*

The category assertion has to be read between the lines of Poincaré's discussion. He began by showing that recurrent points are dense in G. His proof involved the construction of a nested sequence of regions; it may be interpreted as amounting to a proof of Baire's theorem for the case in hand. Since it is a trivial matter to show that the set of points recurrent with respect to G is a G_δ set, the category assertion may properly be ascribed to Poincaré even though he makes no such explicit statement. This part of Poincaré's reasoning was subsequently generalized and extended by G. D. Birkhoff [3, Chapter 7]. The category assertion was made explicit by Hilmy [14].

The measure assertion of Theorem 17.1 was formulated by Poincaré in terms of "probability." In this part of his proof he tacitly assumed the countable additivity of "probability," although this had not been properly justified at the time he was writing. However, when read against an adequate background of measure theory, his argument is perfectly sound. It was reformulated in modern terms by Carathéodory [7].

Closer analysis of Poincaré's reasoning reveals that the assumed preservation of volume is not really essential. In the first part of his reasoning it is used only to exclude the possibility of an open set whose images are mutually disjoint, and in the second part it serves only to exclude the possibility of such a set having positive measure. Moreover, there is no need to assume that T is one-to-one. When stripped of inessential features, both parts of Poincaré's theorem are seen to be contained in a single abstract recurrence theorem, which we shall now formulate and prove.

Let X be a set, let S be a σ-ring of subsets of X, and let I be a σ-ideal in S. A mapping T of X into X is called S-measurable if $T^{-1}E \in S$ whenever $E \in S$. A set $E \subset X$ is called a wandering set if the sets $E, T^{-1}E, T^{-2}E, \ldots$ are mutually disjoint. T is called dissipative if there exists a wandering set that belongs to $S - I$; otherwise T is called nondissipative. For any set $E \subset X$, let $D(E)$ denote the set of points x in E such that $T^i x \in E$ for at most a finite number of positive integers i. T is said to have the recurrence property if $D(E) \in I$ for every $E \in S$.

Theorem 17.2. *An S-measurable mapping T of X into X has the recurrence property if and only if T is nondissipative.*

Proof. Suppose T is nondissipative. Consider any $E \in S$, and let $F = E - \bigcup_1^\infty T^{-k}E$. Since T is S-measurable and S is a σ-ring, F belongs to S. For any integers $0 \leq i < j$ we have

$$T^{-j}F \cap T^{-i}F \subset T^{-j}E - \bigcup_{k=1}^\infty T^{-i-k}E = \emptyset.$$

This shows that F, and each of the sets $T^{-k}F$ $(k = 1, 2, \ldots)$, is a wandering set. Since all of these sets belong to S, and T is nondissipative, it follows that $T^{-k}F \in I$ for all $k \geq 0$. Since I is a σ-ideal, the union $\bigcup_0^\infty T^{-k}F$ belongs to I, and so does the set $H = E \cap \bigcup_0^\infty T^{-k}F$. But $T^{-k}F$ consists of all points x such that $T^k x \in E$ and such that $T^i x \in X - E$ for all $i > k$. Hence $H = D(E)$. Thus we have shown that $D(E) \in I$, for every $E \in S$; that is, T has the recurrence property.

Conversely, if T is dissipative there exists a wandering set E that belongs to $S - I$. Then $D(E) = E$, and we have $E \in S$ but $D(E) \notin I$. This shows that T lacks the recurrence property. []

Both parts of Theorem 17.1 are implied by Theorem 17.2. Suppose first that T is a one-to-one measure-preserving transformation of a bounded open region X of r-space onto itself. Take S to be the σ-algebra of measurable subsets of X, and I to be the σ-ideal of nullsets. Then T is

S-measurable. Since the measure of X is finite, any measurable wandering set must be a nullset, therefore T is nondissipative. Consequently, T has the recurrence property. This means that almost all points of any measurable set E return to E infinitely often under iteration of T. In particular, for any open set $G \subset X$, all points of G except a set of measure zero are recurrent with respect to G.

Next suppose that T is a homeomorphism of a metric space X onto itself, with the property that there is no non-empty open wandering set. (This will be the case if X is a bounded open subset of r-space and T is volume-preserving.) Take S to be the σ-algebra of subsets of X having the property of Baire, and let I be the σ-ideal of sets of first category in X. Then T is S-measurable. By the Banach category theorem, there is a largest open set H of first category. Let $Y = X - \bar{H}$. Then any non-empty open subset of Y is of second category. Evidently H, and therefore Y, is invariant under T. Let E be any wandering set having the property of Baire. Then $E = G \bigtriangleup P$, where G is open and P is of first category. We may assume that $G \subset Y$. For any integers $0 \le i < j$ we have $T^{-i}E \cap T^{-j}E = \emptyset$, which implies that

$$T^{-i}G \cap T^{-j}G = (T^{-i}P \cap T^{-j}G) \bigtriangleup (T^{-i}G \cap T^{-j}P) \bigtriangleup (T^{-i}P \cap T^{-j}P)$$
$$\subset T^{-i}P \cup T^{-j}P .$$

Hence $T^{-i}G \cap T^{-j}G$ is an open subset of Y of first category, therefore empty. Consequently G is an open wandering set, and therefore empty. This shows that any wandering set that has the property of Baire is of first category. Thus T is nondissipative, and so it has the recurrence property. This means that if E has the property of Baire, then all points of E except a set of first category return to E infinitely often under iteration of T. In particular, for any open set $G \subset X$, all points of G except a set of first category are recurrent with respect to G.

Theorem 17.1 is sometimes called the Poincaré recurrence theorem, but more properly this title belongs to the theorem we are about to deduce. First we need another definition. A point x is said to be *recurrent under T* if is recurrent with respect to every neighborhood of itself. (Poincaré called such a point "stable à la Poisson.")

Theorem 17.3 (Poincaré Recurrence Theorem). *If T is a measure-preserving homeomorphism of a bounded open region X of r-space onto itself, then all points of X except a set of first category and measure zero are recurrent under T.*

Proof. Let $U_1, U_2, \ldots$ be a countable base for X. Let E_k be the set of points x in U_k such that $T^i x \in U_k$ for at most a finite number of positive integers i. By Theorem 17.1, each of the sets E_k is a nullset of first category.

Hence the set $E = \bigcup_1^\infty E_k$ is also a nullset of first category. If $x \in X - E$, and if U is any neighborhood of x, then $x \in U_k \subset U$ for some k, and $x \notin E_k$. Hence $T^i x \in U$ for infinitely many positive integers i. Therefore each point of $X - E$ is recurrent under T. □

The significance of this theorem for the theory of dynamical systems rests on the following considerations. In classical mechanics, the configuration of a system is described by a finite set of coordinates $q_1, q_2, \ldots, q_N$. A "state" of the system is specified by the instantaneous values of these coordinates and of the corresponding momenta $p_1, p_2, \ldots, p_N$. These $2N$ values are represented by a point of a $2N$-dimensional space. These points constitute the *phase space* of the system. The points of the phase space represent all possible states of the system. As the state of the system changes in time, in accordance with the equations of motion that govern the system, the representative point describes a path in the phase space. If we follow the motion for unit time, any initial point x in the phase space moves to a point Tx. Thus the equations of motion, followed for unit time, determine a transformation T of phase space into itself. The fundamental existence and uniqueness theorems for solutions of systems of differential equations imply that T is a homeomorphism, provided the terms of the equations are sufficiently continuous and differentiable. Moreover, the Newtonian equations, written in terms of suitable coordinates and momenta (Hamiltonian form), are such that the transformation T preserves $2N$-dimensional measure. This result is known as *Liouville's theorem* (the same Liouville whom we encountered in Chapter 2). For a conservative system, the total energy is constant. Therefore T transforms any surface of constant energy into itself. For some systems, it can be shown that the part of phase space where the energy is suitably restricted is a bounded open region of $2N$-space. Then Theorem 17.3 implies that for almost all initial states (in the sense of either measure or category) the system will return infinitely often arbitrarily close to its initial state. Poisson had attempted to establish this kind of stability in the "restricted problem of three bodies" by an inconclusive argument based on the kind of terms that can appear in certain series expansions. Poincaré established the conclusion rigorously and by a revolutionary new kind of reasoning. This was one of the first triumphs of the modern "qualitative" theory of differential equations, a theory which Poincaré initiated.

Subsequent work on the theory of measure-preserving transformations has shown that Poincaré's theorem can be vastly improved. The "ergodic theorem" of G. D. Birkhoff (1931) asserts that under a measure-preserving transformation of a set of finite measure onto itself, not only do almost all points of any measurable set E return to E infinitely often, but they return with a well-defined positive limiting frequency. More precisely, if χ_E denotes the indicator function of E, then the limiting frequency

$$f(x) = \lim_{n \to \infty} \frac{1}{n} \sum_{i=0}^{n-1} \chi_E(T^i x)$$

exists and is positive for almost all x in E. Evidently this result goes far beyond Theorem 17.1. Curiously, though, this refinement of Poincaré's theorem turns out to be generally false in the sense of category; the set of points where $f(x)$ is defined may be only of first category. The analogy between category and measure goes a long way here, but eventually it breaks down.

18. Transitive Transformations

We have given many illustrations of the category method, but in most cases it has only served to give a new, sometimes simpler, existence proof for objects whose existence was already known. Liouville numbers, nowhere differentiable continuous functions, Brouwer's transformation of the square, were known before the category method was applied. It may therefore be of interest to consider one problem whose solution was first obtained by the category method.

Problem 18.1. To find a homeomorphism T of the closed unit square onto itself such that the positive semiorbit x, Tx, T^2x, ... of some point x is dense in the square.

An automorphism T of a topological space X is said to be *transitive* if there exists a point x whose orbit $\{T^n x : n = 0, \pm 1, \pm 2, ...\}$ is dense in X. When X is a complete separable metric space without isolated points, the existence of such a point implies that points whose positive semiorbit is dense constitute a residual set in X. For if $\{U_i\}$ is a countable base, let

$$G_j = \bigcup_{n=0}^{\infty} T^{-n} U_j \quad (j = 1, 2, ...)$$

and

$$E = \bigcap_{j=1}^{\infty} G_j.$$

Then $x \in E$ if and only if the positive semiorbit of x is dense in X. For any two positive integers i and j, either $T^n U_i \cap U_j \neq \emptyset$ or $T^n U_j \cap U_i \neq \emptyset$ for some integer $n \geq 0$. In the latter case, both x and $T^m x$ belong to $T^n U_j \cap U_i$ for some x and some $m > n$, since every non-empty open set contains infinitely many points of any dense orbit. In either case, $U_i \cap G_j \neq \emptyset$. Hence G_j is a dense open set and E is residual in X. Thus an equivalent statement of Problem 18.1 is this: To find a transitive automorphism of the closed unit square.

Some spaces admit a transitive automorphism and some do not. For example, no automorphism of the unit interval is transitive; but multiplication by $e^{2\pi i \alpha}$, α irrational, defines a transitive rotation of the unit circle in the complex plane. An explicit example of a transitive automorphism of the plane was given by Besicovitch [2], but it is not easy to exhibit one for the closed unit square, let alone one that preserves area or leaves the boundary points fixed. The existence of such transformations was first established by the category method [24]. Similar reasoning shows that there exist transitive automorphisms of any region of Euclidean r-space, $r \geq 2$. The category method has also been successfully used to establish the general existence of automorphisms possessing a much stronger property of the same kind, called *metrical transitivity* [28].

Consider the space H of all automorphisms of the unit square X, with the uniform metric

$$\varrho(S, T) = \sup |Sx - Tx| .$$

H is topologically complete, for the same reason as in the 1-dimensional case (Chapter 13). Suppose that T is transitive and that x is a point whose positive semiorbit is dense. Let $\varepsilon > 0$ be small enough so that the disk with radius ε and center x is contained in X. Let n be the least positive integer such that $|x - T^n x| < \varepsilon$. Choose an open disk U with center x and radius less than ε such that $T^n x$ is interior to U and $Tx, T^2 x, \ldots,$ $T^{n-1} x$ belong to $X - \bar{U}$. Then we can find a closed disk D with center x such that D and $T^n(D)$ are contained in U, and $T(D), T^2(D), \ldots, T^{n-1}(D)$ are contained in $X - U$. Let S be an automorphism of X equal to the identity outside U, and inside U equal to a radial contraction that maps $T^n(D)$ onto a subset of the interior of D. Then $S \circ T$ is an automorphism of X such that $(S \circ T)^n$ maps D onto a subset of its interior. Consequently, $S \circ T$ is not transitive, and neither is any automorphism sufficiently close to it in H. But $\varrho(S \circ T, T) < \varepsilon$. This shows that the transitive automorphisms constitute only a nowhere dense subset of H. Applied to H, Baire's theorem gives no assurance that such transformations exist. The category method appears to have failed!

But suppose we make the problem harder and demand in addition that T preserve measure! Then it is appropriate to consider the space M of measure-preserving automorphisms of the square, with the same metric as before. Since M is a closed subset of H, M is topologically complete.

Let $\{U_i\}$ be an enumeration of all open squares with rational vertices contained in X. For any two positive integers i and j, let

$$E_{ij} = \bigcup_{k=1}^{\infty} \{T \in M : U_i \cap T^{-k} U_j \neq \emptyset\} .$$

It is clear that E_{ij} is open in M. Is it dense? First consider the sets

$$P_{ij} = \{T \in M : T^j x = x \text{ for all } x \text{ in } U_i\}$$

and

$$P = \bigcup_{i,j} P_{ij} .$$

Evidently P_{ij} is closed in M. If T belongs to P_{ij}, then each point of U_i has a period that divides j. Hence we can find a disk D contained in U_i, with arbitrarily small radius, such that for some positive integer k (a divisor of j), the sets $D, T(D), ..., T^{k-1}(D)$ are disjoint and T^k is equal to the identity on D. Let S be a measure-preserving automorphism of X that rotates some annulus concentric with D through an irrational multiple of π and is equal to the identity outside D. Then no point of this annulus is periodic under $S \circ T$. Hence $S \circ T$ belongs to $M - P_{ij}$ and $\varrho(S \circ T, T)$ is arbitrarily small. Consequently, P_{ij} is nowhere dense and P is of first category in M. The set $M - P$ consists of all measure-preserving automorphisms of the square that have nonperiodic points in every non-empty open subset of the square.

For any i and j, any $T \in M - P$, and any $\varepsilon > 0$, construct a transformation S as follows. Join a point of U_i to a point of U_j by a line segment. Choose points $p_1, p_2, ..., p_{N+1}$ on this segment such that $p_1 \in U_i, p_{N+1} \in U_j$, and $|p_k - p_{k+1}| < \varepsilon/2$ $(k = 1, ..., N)$. Choose a positive number $\delta < \frac{1}{2} \min |p_k - p_{k+1}|$ such that the δ-neighborhoods of p_1 and p_{N+1} are contained in U_i and U_j, respectively. Choose a nonperiodic point x_1 in the δ-neighborhood of p_1 that has some point $T^{n_1} x_1$ of its positive semiorbit in the same neighborhood. This is possible, because the recurrent points under T constitute a residual set in the square (by Theorem 17.3), and so do the nonperiodic points, for any T in $M - P$. Similarly, for $k = 2, ..., N + 1$, choose a nonperiodic point x_k in the δ-neighborhood of p_k such that $T^{n_k} x_k$ lies in this neighborhood for some $n_k > 0$. Furthermore, let x_k be chosen so that it does not belong to the orbit of any of the points $x_1, ..., x_{k-1}$. Then all of the points of the set

$$F = \{T^n x_k : 0 \leq n \leq n_k, \ 1 \leq k \leq N + 1\}$$

are distinct, and

$$|T^{n_k} x_k - x_{k+1}| \leq |T^{n_k} x_k - p_k| + |p_k - p_{k+1}| + |p_{k+1} - x_{k+1}|$$
$$< \delta + \varepsilon/2 + \delta < \varepsilon$$

for $k = 1, 2, ..., N$. Hence there exist disjoint open regions $R_1, ..., R_N$, of diameter less than ε, such that

$$R_k \cap F = \{T^{n_k} x_k, x_{k+1}\} \qquad (k = 1, ..., N).$$

(Each R_k may be taken to be a neighborhood of a suitable arc joining $T^{n_k} x_k$ and x_{k+1}.) It is an easy matter to define a transformation S belonging to M that is equal to the identity outside the regions $R_1, ..., R_N$ and that takes $T^{n_k} x_k$ into x_{k+1} $(k = 1, ..., N)$. Then the automorphism

$$(S \circ T)^{n_1 + n_2 + \cdots + n_N}$$

takes x_1 into x_{N+1}. Hence $S \circ T$ belongs to E_{ij}. Since $\varrho(S \circ T, T) < \varepsilon$, it follows that E_{ij} is dense in M. By the Baire category theorem, the set

$\bigcap_{i,j} E_{ij}$ is a dense G_δ in M, and therefore non-empty. For any T in $\bigcap E_{ij}$ we have

$$U_i \cap \bigcup_{k=1}^{\infty} T^{-k} U_j \neq \emptyset$$

for all i and j. Hence the set

$$G_j = \bigcup_{k=1}^{\infty} T^{-k} U_j$$

is open and dense in the square. Therefore $\bigcap G_j \neq \emptyset$ (again by the Baire category theorem!). For any point x in this set, the sequence x, Tx, $T^2 x$, ... is dense in the square. Consequently, any T in $\bigcap E_{ij}$ is transitive.

19. The Sierpinski-Erdös Duality Theorem

The principal use for the notion of category is in the formulation of existence proofs. This may be termed its *telescopic* function. Baire's theorem enables us to bring into focus mathematical objects which otherwise may be difficult to see! But the study of category serves another purpose, too. By developing the theories of measure and category simultaneously, and by calling attention to their points of similarity and difference, we have tried to show how the two theories illuminate each other. Because the theory of measure is more extensive and "important" than that of category, the service is mainly in the direction of measure theory. This may be termed the *stereoscopic* function of the study of category; it adds perspective to measure theory! The suggestion to look for a category analogue, or a measure analogue, has very often proved to be a useful guide. In this and the following chapters we shall take a closer look at the duality we have observed between measure and category, to see how far it extends in the case of the line and other spaces, and to discover what underlies it.

Let us recall some of the similarities between the class of nullsets and the class of sets of first category on the line. Both are σ-ideals. Both include all countable sets. Both include some sets of power c. Both classes have power 2^c (unlike the σ-ideal of countable sets, which has power c). Neither class includes an interval; the complement of any set of either class is dense on the line. Both classes are invariant under translation. Any member of either class is contained in a Borel member of the class.

We have also noted some differences. Any nullset is contained in a G_δ nullset, whereas any set of first category is contained in an F_σ set of first category. Neither class includes the other. The line can be decomposed into a pair of complementary sets, one of first category and the other of measure zero.

Another similarity, which we have not explicitly mentioned, is the following.

Theorem 19.1. *The complement of any linear nullset contains a nullset of power c. The complement of any linear set of first category contains a first category set of power c.*

Proof. By Theorem 3.18, the complement of a nullset contains an uncountable closed set. By Lemma 5.1, this contains a closed nullset of power c.

The complement of a set of first category contains an uncountable G_δ set. By Lemma 5.1, this contains a nowhere dense set of power c. □

The properties of these two σ-ideals suggest that perhaps they are similar, in the following technical sense of this term. A class K of subsets of X is said to be *similar* to a class L of subsets of Y if there exists a one-to-one mapping f of X onto Y such that $f(E) \in L$ if and only if $E \in K$.

In 1934, Sierpinski [34] proved the following

Theorem 19.2 (Sierpinski). *Assuming the continuum hypothesis, there exists a one-to-one mapping f of the line onto itself such that $f(E)$ is a nullset if and only if E is of first category.*

This theorem explains many of the similarities we have noted. In fact, it justifies the following principle of duality: Let P be any proposition involving solely the notion of nullset and notions of pure set theory (for example, cardinality, disjointness, or any property invariant under arbitrary one-to-one transformation). Let P^* be the proposition obtained from P by replacing "nullset" by "set of first category" throughout. Then each of the propositions P and P^* implies the other, assuming the continuum hypothesis. It is not known whether Sierpinski's theorem can be proved without the continuum hypothesis.

Sierpinski asked whether a stronger theorem may be true: Does there exist a mapping f that maps each of the two classes onto the other simultaneously? This question was answered in 1943 by Erdös [10]. By a relatively small refinement of Sierpinski's proof, Erdös proved the following

Theorem 19.3 (Erdös). *Assuming the continuum hypothesis, there exists a one-to-one mapping f of the line onto itself such that $f = f^{-1}$ and such that $f(E)$ is a nullset if and only if E is of first category. (It follows from these properties that $f(E)$ is of first category if and only if E is a nullset.)*

The interest of this theorem is that it establishes a stronger form of duality, which may be stated as follows.

Theorem 19.4 (Duality Principle). *Let P be any proposition involving solely the notions of measure zero, first category, and notions of pure set theory. Let P^* be the proposition obtained from P by interchanging the terms "nullset" and "set of first category" wherever they appear. Then each of the propositions P and P^* implies the other, assuming the continuum hypothesis.*

We shall base our proof of the Sierpinski-Erdös theorem on the following purely set-theoretic

Theorem 19.5. *Let X be a set of power $\aleph_1$, and let K be a class of subsets of X with the following properties:*

(a) *K is a σ-ideal,*

(b) *the union of K is X,*

(c) *K has a subclass G of power $\leq \aleph_1$ with the property that each member of K is contained in some member of G,*

(d) *the complement of each member of K contains a set of power $\aleph_1$ that belongs to K.*

Then X can be decomposed into $\aleph_1$ disjoint sets X_α, each of power $\aleph_1$, such that a subset E of X belongs to K if and only if E is contained in a countable union of the sets X_α.

Proof. Let $A = \{\alpha : 0 \leq \alpha < \Omega\}$ be the set of ordinals of first or second class, that is, all ordinals less than the first ordinal, Ω, that has uncountably many predecessors. Then A has power $\aleph_1$, and there exists a mapping $\alpha \to G_\alpha$ of A onto G. For each $\alpha \in A$ define

$$H_\alpha = \bigcup_{\beta \leq \alpha} G_\beta \quad \text{and} \quad K_\alpha = H_\alpha - \bigcup_{\beta < \alpha} H_\beta .$$

Put $B = \{\alpha \in A : K_\alpha \text{ is uncountable}\}$. Properties (a), (c) and (d) imply that B has no upper bound in A. Therefore there exists a one-to-one order-preserving map ϕ of A onto B. For each α in A, define

$$X_\alpha = H_{\phi(\alpha)} - \bigcup_{\beta < \alpha} H_{\phi(\beta)} .$$

By construction and property (a), the sets X_α are disjoint and belong to K. Since $X_\alpha \supset K_{\phi(\alpha)}$, each of the sets X_α has power $\aleph_1$. For any $\beta \in A$, we have $\beta < \phi(\alpha)$ for some $\alpha \in A$, and therefore

$$G_\beta \subset H_\beta \subset H_{\phi(\alpha)} = \bigcup_{\gamma \leq \alpha} X_\gamma .$$

Hence, by (c), each member of K is contained in a countable union of the sets X_α. Using (b), it follows that

$$X = \bigcup K \subset \bigcup_{\alpha \in A} X_\alpha .$$

Thus $\{X_\alpha : \alpha \in A\}$ is a decomposition of X with the required properties. $\square$

Theorem 19.6. *Let X be a set of power $\aleph_1$. Let K and L be two classes of subsets of X each of which has properties (a) to (d) of Theorem 19.5. Suppose further that X is the union of two complementary sets M and N, with $M \in K$ and $N \in L$. Then there exists a one-to-one mapping f of X onto itself such that $f = f^{-1}$ and such that $f(E) \in L$ if and only if $E \in K$.*

Proof. Let X_α $(0 \leq \alpha < \Omega)$ be a decomposition of X corresponding to K, as constructed in the proof of Theorem 19.5. We may assume that M belongs to the generating class G, and that G_0 is taken equal to M. Then $X_0 = M$, because M cannot be countable. Similarly, let Y_α $(0 \leq \alpha < \Omega)$ be a decomposition of X corresponding to L, with $Y_0 = N$. Then

$$M = \bigcup_{0 < \alpha < \Omega} Y_\alpha \quad \text{and} \quad N = \bigcup_{0 < \alpha < \Omega} X_\alpha.$$

The sets X_α and Y_α, for $0 < \alpha < \Omega$, constitute a decomposition of X into sets of power $\aleph_1$. For each $0 < \alpha < \Omega$, let f_α be a one-to-one mapping of X_α onto Y_α. Define f equal to f_α on X_α, and equal to f_α^{-1} on Y_α, for $0 < \alpha < \Omega$. Then f is a one-to-one mapping of X onto itself, f is equal to f^{-1}, and $f(X_\alpha) = Y_\alpha$ for all $0 < \alpha < \Omega$. Since

$$X_0 = \bigcup_{0 < \alpha < \Omega} Y_\alpha \quad \text{and} \quad Y_0 = \bigcup_{0 < \alpha < \Omega} X_\alpha,$$

we have also $f(X_0) = Y_0$. Thus $f(X_\alpha) = Y_\alpha$ for all $0 \leq \alpha < \Omega$. From the properties of X_α and Y_α stated in Theorem 19.5 it follows that $f(E) \in L$ if and only if $E \in K$. □

Theorem 19.3 is an immediate consequence of Theorem 19.6. Take X to be the line, let K be the class of sets of first category, and let L be the class of nullsets. K is generated by the class of F_σ sets of first category, and L by the class of G_δ nullsets. Each of these generating classes has power c. Condition (c) is therefore satisfied, on the hypothesis that $c = \aleph_1$. Condition (d) is implied by Theorem 19.1, and conditions (a) and (b) are obvious. For the sets M and N we may take the sets A and B of Theorem 1.6.

20. Examples of Duality

In 1934, in his book entitled "Hypothèse du continu" [35], Sierpinski collected a number of examples of dual propositions (in the restricted sense, since Erdös's theorem had not yet been proved). Here we shall discuss three of these pairs, in some cases modifying the statement slightly. In each case, the propositions have been proved up to now only with the aid of the continuum hypothesis. Hence the duals follow without loss from the duality principle. Because they depend on the continuum hypothesis, we shall designate them as "propositions" rather than as "theorems."

The first proposition is due to Lusin (1914) [35, pp. 36, 81).

Proposition 20.1. *Any linear set E of second category has a subset N of power c such that every uncountable subset of N is of second category.*

Proof. Let $\{X_\alpha : \alpha < \Omega\}$ be the decomposition of X corresponding to the class K of first category sets in the proof of Theorem 19.5. Let N be a set obtained by selecting just one point from each non-empty set of the form $E \cap X_\alpha$. Since E is of second category, N is uncountable and therefore of power c. No uncountable subset of N can be covered by countably many of the sets X_α. Hence no uncountable subset of N is of first category. $\square$

An uncountable set with the property that every uncountable subset is of second category is called a *Lusin set*. The dual proposition was proved by Sierpinski in 1924 [35, pp. 80, 82].

Proposition 20.1*. *Any linear set E of positive outer measure has a subset N of power c such that every uncountable subset of N has positive outer measure.*

In Proposition 20.1, with $E = R$, we can adjoin to N a countable dense set and thereby insure that N is dense. Then a subset of N is of first category relative to N if and only if it is of first category in R. Hence these propositions imply that there are uncountable subspaces of the line in which the distinction between first and second category, or between

Lebesgue nullset or not, reduces to the distinction between countable and uncountable.

Since every subset of the line is the union of a nullset and a set of first category (Corollary 1.7), a Lusin set must have measure zero and the set N of Proposition 20.1* must be of first category.

Proposition 20.2. *There exists a one-to-one mapping f of the line onto a subset of itself such that $f(E)$ is of second category whenever E is uncountable.*

Proof. Let f be any one-to-one mapping of the line onto a Lusin set. ⬜

Proposition 20.2*. *There exists a one-to-one mapping f of the line onto a subset of itself such that $f(E)$ has positive outer measure whenever E is uncountable.*

Proposition 20.3. *Any linear set E of second category contains c disjoint subsets of second category.*

Proof. Let f be a one-to-one mapping of R onto a Lusin set contained in E. Then the conclusion follows from the fact that the line and plane have the same cardinality. ⬜

Proposition 20.3*. *Any linear set E of positive outer measure contains c disjoint sets of positive outer measure.*

As we showed earlier (Theorem 5.5), any set with positive outer measure contains a non-measurable subset. Hence Proposition 20.3* implies that any set E of positive measure contains c disjoint non-measurable subsets. By Zorn's lemma, this family is contained in a maximal disjoint class of non-measurable subsets of E. The complement of the union of such a family must have measure zero. By adjoining it to one of the members of the family we obtain a decomposition of E into c disjoint non-measurable subsets. In this connection it is interesting to note that Lusin and Sierpinski [36] have proved, without assuming the continuum hypothesis, that the line can be decomposed into c disjoint Bernstein sets, and therefore into c disjoint sets of positive outer measure. At the same time, this construction establishes the dual proposition – that the line can be decomposed into c disjoint sets of second category. It can be shown that these properties imply that the algebra of all subsets of the line modulo the ideal of nullsets, or modulo the ideal of sets of first category, is incomplete. This means that not every subset of the quotient algebra has a least upper bound [37, footnote 11].

None of the dual propositions we have so far considered has involved measure and category simultaneously. Let us now consider some examples of duality in the more general sense. An obvious instance is

Theorem 1.6: The line can be decomposed into two complementary sets, one of first category, the other of measure zero. This proposition is self-dual. A slightly more interesting example is this: A subset of R is a nullset if its intersection with every set of first category is countable. The dual reads: A subset of R is of first category if its intersection with every nullset is countable. Both of these are corollaries of Theorem 1.6; the continuum hypothesis is not needed.

Let us now consider a nontrivial example. It is a slight generalization of one of Sierpinski's propositions [35, p. 130].

Proposition 20.4. *For any class K of one-to-one nullset-preserving transformations of the line, with card $K = c$, there exists a linear set E of first category and power c such that $TE \vartriangle E$ is a countable set, for each T in K.*

Proof. Index the elements of K and R so that

$$K = \{T_\alpha : \alpha < \Omega\} \quad \text{and} \quad R = \{p_\alpha : \alpha < \Omega\}.$$

Let A be a nullset such that $R - A$ is of first category. For $0 < \alpha < \Omega$, let G_α be the group generated by the transformations T_β with $\beta < \alpha$. Then G_α consists of all products of the form

$$T_{\beta_1}^{k_1} T_{\beta_2}^{k_2} \dots T_{\beta_n}^{k_n},$$

where $\beta_i < \alpha$ and $k_i = \pm 1$ $(i = 1, \dots, n)$, and n is any positive integer. Hence G_α is countable, and each T in G_α is nullset-preserving. For each T in G_α, the set TA is a nullset. Hence $A_\alpha = \bigcup \{TA : T \in G_\alpha\}$ is a nullset. Let $x_0 = p_0$. Assuming that points x_β in R have been defined for all $\beta < \alpha$, put $B_\alpha = \{Tx_\beta : \beta < \alpha, \; T \in G_\alpha\}$. Then B_α is a countable set, and $A_\alpha \cup B_\alpha$ is a nullset. Let x_α be the first element in the well ordering of R such that x_α is not in $A_\alpha \cup B_\alpha$. Put $E_\alpha = \{Tx_\alpha : T \in G_\alpha\}$, and define $E = \bigcup_{0 < \alpha < \Omega} E_\alpha$. Then E_α is countable, and E is uncountable. Moreover, E is a subset of $R - A$. Hence E is of first category. For any $\beta < \alpha < \Omega$, we have $T_\beta E_\alpha = E_\alpha$. Hence $T_\beta E \vartriangle E \subset \bigcup_{\alpha \leq \beta} (E_\alpha \cup T_\beta E_\alpha)$. This shows that $TE \vartriangle E$ is countable, for each T in K. $\Box$

Proposition 20.4*. *For any class K of one-to-one category-preserving transformations of the line, with card $K = c$, there exists a linear set E of measure zero and power c such that $TE \vartriangle E$ is a countable set, for each T in K.*

Sierpinski proved these propositions for the class K of translations. However, the notion of a translation is not purely set-theoretic, and so the corresponding propositions are not strictly dual. The class of all homeomorphisms of the line onto itself also has power c, and we can deduce from Proposition 20.4* the following

Corollary 20.5. *Assuming the continuum hypothesis, there exists an uncountable linear set E such that the image of E under any automorphism of the line is a nullset.*

This shows that in Theorem 13.2, the hypothesis that A be a closed set cannot be omitted (though it can be replaced by a weaker hypothesis, such as that A be a Borel set). Actually, Sierpinski [33, p. 274] has shown that an even stronger form of the above corollary can be proved without assuming the continuum hypothesis. But, curiously enough, it has not been proved that such a set can have power c. Assuming the continuum hypothesis, we have seen that not only does such a set exist, but it can be chosen in such a way as to differ from its image under any automorphism of the line by only a countable set.

21. The Extended Principle of Duality

We have seen many instances in which the property of Baire has played a role analogous to measurability. A typical example is the following, which is a restatement of Theorem 5.5.

Theorem 21.1. *If every subset of a linear set E is measurable, then E is a nullset. If every subset of E has the property of Baire, then E is of first category.*

Can the principle of duality be extended to include measurability and the property of Baire as dual notions? The possibility of such a principle was first considered by Szpilrajn [38]. To justify it, we should like to find a one-to-one mapping f of the line onto itself such that $f(E)$ is measurable if and only if E has the property of Baire, and such that $f(E)$ is a nullset if and only if E is of first category. (The second property is a consequence of the first, by the above theorem and its converse.) However, as Szpilrajn showed, such a mapping is impossible.

For suppose f is such a mapping. Let I be the unit interval, and let $E = f^{-1}(I)$. Then E has the property of Baire. Let $x_1, x_2, \ldots$ be a countable dense subset of E, and let I_i be an open interval containing x_i such that

$$m(f(I_i) \cap I) < 1/2^{i+1}.$$

Put $G = \bigcup I_i$. Then G is an open set and $E \subset \bar{G}$. Hence

$$E \subset (G \cap E) \cup (\bar{G} - G).$$

Therefore

$$I = f(E) \subset f(G \cap E) \cup f(\bar{G} - G) \subset \bigcup [f(I_i) \cap I] \cup f(\bar{G} - G).$$

Since $\bar{G} - G$ is nowhere dense, $f(\bar{G} - G)$ is a nullset, and so $m(I) \leq \sum_1^\infty 2^{-i-1} = 1/2$, a contradiction.

The foregoing argument shows, incidentally, that in Theorem 19.2 the mappings f and f^{-1} cannot both be Borel measurable, that is, we cannot require that $f(E)$ and $f^{-1}(E)$ be Borel sets whenever E is a Borel set. (In fact, neither f nor f^{-1} can be Borel measurable. This is because the inverse of a one-to-one Borel measurable mapping is also Borel measurable [18, p. 398].)

To clinch the conclusion that the extended principle of duality is not valid as a general principle, let us examine an instance in which it breaks down.

Theorem 21.2. *Let $E_{i,j}$ be a double sequence of measurable sets such that $E_{i,j} \supset E_{i,j+1}$ for all positive integers i and j, and such that $\bigcap_j E_{i,j}$ is a nullset for each i. Then there exists a sequence of mappings $n_k(i)$ of the set of positive integers into itself such that $\bigcap_k \bigcup_i E_{i,n_k(i)}$ is a nullset.*

Proof. Let $I_k = [-k, k]$. For each i and k there is a positive integer $n_k(i)$ such that

$$m(E_{i,n_k(i)} \cap I_k) < 1/k2^i .$$

Hence

$$m\left(\bigcup_i E_{i,n_k(i)} \cap I_k\right) < 1/k .$$

Put $E = \bigcap_k \bigcup_i E_{i,n_k(i)}$. For any finite interval I, we have $I \subset I_k$ for all sufficiently large k. Then

$$(E \cap I) \subset \bigcup_i E_{i,n_k(i)} \cap I_k .$$

Hence $m(E \cap I) < 1/k$ for all sufficiently large k. Thus $E \cap I$ is a nullset for every I, and therefore E is a nullset. $\square$

Dual Statement. *Let $E_{i,j}$ be a double sequence of sets having the property of Baire such that $E_{i,j} \supset E_{i,j+1}$ for all positive integers i and j, and such that $\bigcap_j E_{i,j}$ is of first category for each i. Then there exists a sequence of mappings $n_k(i)$ of the set of positive integers into itself such that $\bigcap_k \bigcup_i E_{i,n_k(i)}$ is of first category.*

This proposition is false. Let r_i be an enumeration of all rational points, and let $E_{i,j} = (r_i - 1/j, r_i + 1/j)$. This double sequence satisfies the hypothesis of the proposition in question. For any mapping $n(i)$ of the positive integers into positive integers, the set $\bigcup_i E_{i,n(i)}$ is a dense open set. For any sequence of such mappings $n_k(i)$, the set $\bigcap_k \bigcup_i E_{i,n_k(i)}$ is residual, contrary to the stated conclusion.

Although the extended principle of duality is not valid as a general principle, it has a certain heuristic value. Many properties of measure depend only on properties of the class of measurable sets that are shared by the class of sets having the property of Baire. In such cases the principle may suggest (even though it cannot prove) a valid dual. Then one is led to seek an abstract theorem that includes both. Our discussion of the Poincaré recurrence theorem provides an illustration.

The Fubini and Kuratowski-Ulam theorems push the analogy a step further, making the topology of the product space correspond to the product measure. Here the "duality" becomes still more tenuous. Although the statements of the two theorems are strikingly similar, the proofs bear little resemblance to one another, and we were unable to find a generalization that would include both. (The connection we established was of a different nature.)

Can one push the analogy even further, to infinite products? If $\{X_i\}$ is a sequence of sets, the Cartesian product $X = \mathscr{P} X_i$ is the set of all sequences $\{x_i\}$ with $x_i \in X_i$, that is, all functions x from the positive integers to $\bigcup X_i$ such that $x_i \in X_i$ for every i. If the sets X_i are topological or metric spaces, they determine a corresponding topology or metric in X. If the X_i are normalized measure spaces they determine a normalized "product measure" in X. We shall not give a general discussion of these notions here, but confine attention to a particular case.

Let X_i consist of the two elements 0 and 1. Then X is the set of all sequences of zeros and ones. The mapping of X onto the Cantor set defined by

$$f(x) = \sum_{i=1}^{\infty} 2x_i/3^i$$

may be taken to define the product topology in X. Similarly, the mapping

$$g(x) = \sum_{i=1}^{\infty} x_i/2^i$$

of X onto $[0, 1]$, although not one-to-one, defines a measure $\mu(E) = m(g(E))$ on the class of sets E such that $g(E)$ is measurable. This may be taken as the definition of the product measure in X.

A subset E of X is called a *tail set* if whenever $x \in E$ and y differs from x in only a finite number of coordinates, then $y \in E$. Thus, membership in a tail set depends only on the "tail" of a sequence $\{x_i\}$. The notion can be expressed more conveniently as follows. Let

$$X^n = \mathscr{P}_1^n X_i \quad \text{and} \quad Y^n = \mathscr{P}_{n+1}^{\infty} X_i .$$

Then $X = X^n \times Y^n$ for each n. A set $E \subset X$ is a tail set if and only if, for each n, E can be "factored" in the form $E = X^n \times B_n$, where B_n is some subset of Y^n.

An important theorem concerning product measure is the *zero-one law* of Kolmogoroff. For the product space X that we are considering, it reads as follows.

Theorem 21.3. *If E is a measurable tail set in X, then either $\mu(E) = 0$ or $\mu(E) = 1$.*

We merely sketch the proof, specializing that of Halmos [12, p. 201]. Let A_n be a subset of X^n, and put $F = A_n \times Y^n$. Let $E = X^n \times B_n$, where $B_n \subset Y^n$, for each n. Then $E \cap F = A_n \times B_n$. In our case, X^n is a finite set. If A_n has k points, then $\mu(F) = k/2^n$ and $\mu(A_n \times B_n) = (k/2^n)\mu(X^n \times B_n)$. Hence $\mu(E \cap F) = \mu(E)\mu(F)$. Every measurable set can be approximated in the space of measurable sets (Chapter 10) by a set of the form F, since

$g(F)$ can be any finite union of dyadic subintervals of $[0,1]$. It follows that the equation $\mu(E \cap F) = \mu(E)\,\mu(F)$ holds for every measurable set F. In particular, it holds when $F = E$. Hence $\mu(E) = 0$ or 1.

Does this theorem have a category analogue? If so, it should read as follows.

Theorem 21.4. *If E is a tail set in X having the property of Baire, then E is either of first category or residual.*

This theorem is true! For suppose that E is not residual. Then $X - E$ is of the form $G \bigtriangleup P$, G open and non-empty, P of first category. G is a countable union of basic open sets of the form $U = A_n \times Y^n$ (corresponding to the closed intervals used to define the Cantor set). Hence G contains a set U of the form $U = A_n \times Y^n$, where A_n is non-empty. By hypothesis, E can be written in the form $E = X^n \times B_n$. Hence $U \cap E = A_n \times B_n$. But also,

$$A_n \times B_n \subset G \subset (X - E) \cup P.$$

Hence

$$A_n \times B_n \subset E \cap [(X - E) \cup P] \subset P.$$

Therefore $A_n \times B_n$ is of first category. Since A_n is a non-empty subset of the finite discrete space X^n, it is of second category. It follows from Theorem 15.3 that B_n is of first category in Y^n. Hence $E = X^n \times B_n$ is of first category in X.

Although this proof is limited to the particular product space considered, the theorem can be shown to hold for the product of any family of Baire spaces each of which has a countable base [26].

To illustrate Theorem 21.4, let E be the set of sequences $\{x_i\}$ such that $\lim_{n \to \infty} \frac{1}{n} \sum_1^n x_i = 1/2$. This is evidently a tail set in X. It is also a Borel set. It follows that $\mu(E) = 0$ or $\mu(E) = 1$, and that either E or $X - E$ is of first category. Which is it? It is not hard to show (we omit the proof) that E is of first category. On the other hand, the Borel strong law of large numbers implies that $\mu(E) = 1$. This result may be interpreted to mean that the category analogue of the strong law of large numbers is false. Thus it appears that the analogy between measure and category extends through the zero-one law but not as far as the law of large numbers. This is reminiscent of the fact that the analogy extends through the Poincaré recurrence theorem but not as far as the ergodic theorem. The law of large numbers is, in fact, deducible from the ergodic theorem, so these two cases in which the analogy breaks down are not unrelated. Unfortunately, no general criterion for recognizing when a measure theorem has a valid category analogue is known.

22. Category Measure Spaces

It was shown in the last chapter that no one-to-one mapping of the line onto itself can map the nullsets onto the sets of first category and at the same time map the measurable sets onto the sets having the property of Baire. More generally, if (X, S, μ) is a measure space with $0 < \mu(X) < \infty$, and Y is a separable metric space without isolated points, then no S-measurable mapping of X into Y can be such that the inverse image of every set of first category has measure zero. For if f were such a mapping we could define a finite nonatomic Borel measure v in Y by setting $v(E) = \mu(f^{-1}(E))$ for every Borel set E. By Theorem 16.5, Y could be decomposed into a v-nullset and a set of first category. Then $v(Y)$ would be 0, contrary to $\mu(X) > 0$.

However, the possibility remains that in more general topological spaces such a mapping may be possible and no such decomposition exist. This is indeed the case, but we shall see that such spaces have unusual topological properties. Here we shall confine attention to *regular* spaces, that is, Hausdorff spaces in which every neighborhood of a point contains a closed neighborhood of the point. Every compact Hausdorff space is a regular Baire space, and every subspace of a regular space is regular.

If X is a topological space with a finite measure μ defined on the σ-algebra S of sets having the property of Baire, and if $\mu(E) = 0$ if and only if E is of first category, then (X, S, μ) is called a *category measure space* and μ is called a *category measure* in X. In any such space the extended principle of duality is not only valid, it is a tautology! Before discussing the existence of category measures let us determine some of their properties.

Theorem 22.1. *Let μ be a category measure in a regular Baire space X. For any open set G and $\varepsilon > 0$ there is a closed set F such that $F \subset G$ and $\mu(F) > \mu(G) - \varepsilon$, and for every closed set F there is an open set G such that $F \subset G$ and $\mu(G) < \mu(F) + \varepsilon$.*

Proof. Let $\mathscr{F}$ be a maximal disjoint family of non-empty open sets U such that $\bar{U} \subset G$. Each member of $\mathscr{F}$ has positive measure, hence $\mathscr{F}$ is countable, say $\mathscr{F} = \{U_i\}$. Put $U = \bigcup U_i$. Then $U \subset G$. The maximality of $\mathcal{F}$ implies that $G \subset \bar{U}$. Hence $G - U$, which is contained in $\bar{U} - U$, is nowhere dense, and so $\mu(G) = \sum \mu(U_i)$. Choose n so that $\sum_1^n \mu(U_i) > \mu(G) - \varepsilon$. Then $F = \bigcup_1^n \bar{U}_i$ is a closed subset of G, and $\mu(F) > \mu(G) - \varepsilon$. This proves the first assertion; the second follows by complementation. □

Theorem 22.2. *If X is a regular Baire space and μ is a category measure in X, then every set of first category in X is nowhere dense.*

Proof. Let $P = \bigcup N_i$, N_i nowhere dense, be any set of first category. Since $\mu(\bar{N}_i) = 0$, Theorem 22.1 implies that for any two positive integers i and j there is an open set G_{ij} such that $\bar{N}_i \subset G_{ij}$ and $\mu(G_{ij}) < 1/2^{i+j}$. Put $H_j = \bigcup_{i=1}^{\infty} G_{ij}$. Then H_j is open, $P \subset H_j$, and $\mu(\bar{H}_j) = \mu(H_j) < 1/2^j$. Put $F = \bigcap_{j=1}^{\infty} \bar{H}_j$. Then F is closed and $P \subset F$. Since $\mu(F) = 0$, the interior of F must be empty. Hence F, and therefore P, is nowhere dense. □

Theorem 22.3. *If μ is a category measure in a regular Baire space X, then for any set E having the property of Baire,*

$$\mu(E) = \mu(\bar{E}) = \mu(E'^{-\prime})$$

and

$$\mu(E) = \begin{cases} \inf \{\mu(G) : E \subset G, G \text{ open}\} \\ \sup \{\mu(F) : E \supset F, F \text{ closed}\} . \end{cases}$$

Proof. Let $E = G \mathbin{\triangle} P$, G open and P of first category. Then P is nowhere dense, and so is $\bar{P}$. Since

$$G - \bar{P} \subset E \subset G \cup P,$$

we have

$$G - \bar{P} \subset E'^{-\prime} \subset E \subset \bar{E} \subset \bar{G} \cup \bar{P}.$$

The first and last of these sets differ by a nowhere dense set, hence all have equal measure. This proves the first assertion; the second then follows from Theorem 22.1. □

Theorem 22.2 shows that spaces that admit a category measure are topologically unusual. Theorem 22.3 shows that category measures are very tightly fitted to the topology.

We now consider the following problem: Given a finite measure space (X, S, μ), can we define a topology $\mathcal{T}$ in X with respect to which μ will be a category measure? It is obviously necessary to assume that μ is complete, since the class $\mathcal{N}$ of nullsets must be identified with the class of sets of first category. By Theorem 4.5, any open set is of the form $H - \bar{N}$, where H is regular open and N is nowhere dense. A topology is therefore determined by its regular open sets and its nowhere dense closed sets. In a Baire space, any set E belonging to the class S of sets having the property of Baire has a unique representation in the form

$G \bigtriangleup P$, where G is regular open and P is of first category (Theorem 4.6). If we write $G = \phi(E)$, then ϕ is a function that selects a representative element from each equivalence class of S modulo sets of first category. Theorem 4.7 implies that ϕ satisfies conditions similar to those that characterize Lebesgue lower density (Theorem 3.21). This suggests the following program for making a measure space (X, S, μ) into a category measure space. Find a mapping ϕ of S into S that satisfies conditions 1) to 5) of Theorem 3.21. Then find a suitable subclass of $\mathcal{N}$ to serve as the nowhere dense closed sets. We shall show that the class $\mathcal{N}$ itself can always be taken for this purpose. We thereby obtain a maximal topology corresponding to ϕ. To be able to apply this method, we need the following theorem, due to von Neumann and Maharam [19]. Another proof has been given by A. and C. Ionescu Tulcea [15].

Theorem 22.4. *Given a complete finite measure space (X, S, μ), there exists a mapping ϕ of S into S having the following properties, where $A \sim B$ signifies that $A \bigtriangleup B$ belongs to the class $\mathcal{N}$ of μ-nullsets:*
1) $\phi(A) \sim A$,
2) $A \sim B$ *implies* $\phi(A) = \phi(B)$,
3) $\phi(\emptyset) = \emptyset$, $\phi(X) = X$,
4) $\phi(A \cap B) = \phi(A) \cap \phi(B)$,
5) $A \subset B$ *implies* $\phi(A) \subset \phi(B)$.

Such a mapping ϕ is called a *lower density*. We shall not prove this theorem in general. We are primarily interested in the Lebesgue measure space, and we have already seen that in this case the Lebesgue density theorem defines such a mapping (Theorem 3.21). However, it is just as easy to introduce the corresponding topology in the general case. Accordingly, let us assume that (X, S, μ) is a complete finite measure space and that we are given a mapping $\phi : S \to S$ that satisfies conditions 1) to 5). Let $\mathcal{N}$ be the class of μ-nullsets, and define

$$\mathcal{T} = \{\phi(A) - N : A \in S, N \in \mathcal{N}\}.$$

Theorem 22.5. $\mathcal{T}$ *is a topology in X.*

Proof. Since $\emptyset \in \mathcal{N}$, Property 3) implies that $X = \phi(X) - \emptyset$ and $\emptyset = \phi(\emptyset) - \emptyset$ both belong to $\mathcal{T}$. By 4) we have

$$[\phi(A_1) - N_1] \cap [\phi(A_2) - N_2] = \phi(A_1 \cap A_2) - (N_1 \cup N_2).$$

Hence $\mathcal{T}$ is closed under intersection. To show that $\mathcal{T}$ is closed under arbitrary union, let

$$\mathcal{F} = \{\phi(A_\alpha) - N_\alpha : \alpha \in \Gamma\}, \quad A_\alpha \in S, N_\alpha \in \mathcal{N},$$

be any subfamily of $\mathcal{T}$. Let b denote the least upper bound of the measures of finite unions of members of $\mathcal{F}$, and choose a sequence $\{\alpha_n\}$ such that $\mu(\bigcup_1^\infty A_{\alpha_n}) = b$. Put $A = \bigcup_1^\infty A_{\alpha_n}$. Then $A \in S$, and the definition of b implies that $A_\alpha - A \in \mathcal{N}$ for every $\alpha \in \Gamma$. Since

$$A_\alpha - (A_\alpha - A) \subset A ,$$

it follows from 2) and 5) that

$$\phi(A_\alpha) \subset \phi(A) \quad \text{for every } \alpha .$$

Putting

$$N_0 = \bigcup_{n=1}^\infty [N_{\alpha_n} \cup (A_{\alpha_n} - \phi(A_{\alpha_n}))] ,$$

we have $N_0 \in \mathcal{N}$ and

$$A - N_0 \subset \bigcup_{n=1}^\infty [\phi(A_{\alpha_n}) - N_{\alpha_n}] \subset \bigcup_{\alpha \in \Gamma} [\phi(A_\alpha) - N_\alpha] \subset \phi(A) .$$

The extremes differ by a nullset, and therefore

$$\bigcup_{\alpha \in \Gamma} [\phi(A_\alpha) - N_\alpha] = \phi(A) - N$$

for some $N \in \mathcal{N}$, by the completeness of μ. []

This topology has been studied particularly by A. and C. Ionescu Tulcea [16, Chapter 5].

Theorem 22.6. *A set $N \subset X$ is nowhere dense relative to $\mathcal{T}$ if and only if $N \in \mathcal{N}$. Every nowhere dense set is closed.*

Proof. If $N \in \mathcal{N}$, then $X - N = \phi(X) - N \in \mathcal{T}$, hence each member of $\mathcal{N}$ is closed. If $N \in \mathcal{N}$ and $\phi(A_1) - N_1 \subset N$ for some $A_1 \in S$ and $N_1 \in \mathcal{N}$, then $\phi(A_1) \in \mathcal{N}$ and so $\phi(A_1) = \emptyset$, by 2) and 3). Hence $\phi(A_1) - N_1 = \emptyset$, and therefore N is nowhere dense. Conversely, if F is closed and nowhere dense, then $X - F = \phi(A) - N$ for some $A \in S$ and $N \in \mathcal{N}$, hence F belongs to S. Since

$$F \supset \phi(F) - [\phi(F) - F] \in \mathcal{T} ,$$

the nowhere denseness of F implies that $\phi(F) \subset \phi(F) - F$. Hence $\phi(F) = \emptyset$, by 1), 2), and 3). Therefore $F \sim \emptyset$, that is, $F \in \mathcal{N}$. Thus, $\mathcal{N}$ is identical with the class of closed nowhere dense sets. Since every nowhere dense set is contained in a closed nowhere dense set, and every subset of a member of $\mathcal{N}$ belongs to $\mathcal{N}$, it follows that every nowhere dense set is closed. []

Theorem 22.7. *A set $A \subset X$ has the property of Baire if and only if $A \in S$.*

Proof. If $A \in S$, then $A = \phi(A) \bigtriangleup (\phi(A) \bigtriangleup A)$. Since $\phi(A) \in \mathcal{T}$, and $\phi(A) \bigtriangleup A \in \mathcal{N}$, it follows from Theorem 22.6 that A has the property of

Baire. Conversely, if A has the property of Baire, then $A = [\phi(B) - N] \triangle M$ for some $B \in S$, some $N \in \mathcal{N}$, and some set M of first category. By Theorem 22.6, M belongs to $\mathcal{N}$, and therefore $A \in S$. ☐

Theorem 22.8. *A set $G \subset X$ is regular open if and only if $G = \phi(A)$ for some $A \in S$.*

Proof. If $A \in S$, then $\phi(A)$ is open, and the closure of $\phi(A)$ is of the form $\phi(A) \cup N$ for some $N \in \mathcal{N}$, by Theorem 22.6. Let $\phi(A_1) - N_1$ be any open subset of $\phi(A) \cup N$. Then

$$\phi(A_1) - N_1 \subset \phi(A_1) = \phi(\phi(A_1) - N_1) \subset \phi(\phi(A) \cup N) = \phi(A).$$

Thus $\phi(A)$ is the largest open subset of $\phi(A) \cup N$. This shows that $\phi(A)$ is equal to the interior of its closure, that is, $\phi(A)$ is regular open. Conversely, if G is regular open, then $G = \phi(A) - N$ for some $A \in S$ and $N \in \mathcal{N}$. Since $\phi(A) \triangle [\phi(A) - N]$ is contained in N, we have $\phi(A) \sim \phi(A) - N = G$. Since G and $\phi(A)$ differ by a nowhere dense set, and both are regular open, it follows that $G = \phi(A)$. ☐

We have now shown that the problem of topologizing a complete finite measure space (X, S, μ) so as to make it a category measure space is reducible to that of finding a lower density ϕ. In general one can say little about the regularity of the topology $\mathcal{T}$; it need not even be Hausdorff, since S need not separate points of X. However, in the case of Lebesgue measure in R, or in any open interval, we can take $\phi(A)$ to be the set of points where A has density 1. The corresponding topology $\mathcal{T}$ is called the *density topology*. $\mathcal{T}$ consists of all measurable sets A such that A has density 1 at each of its points. Hence $\mathcal{T}$ includes all sets that are open in the ordinary topology, consequently it is Hausdorff. In fact, the density topology in R can be shown to be completely regular but not normal [11]. We shall show only that it is regular.

Theorem 22.9. *The density topology in R is regular.*

Proof. Let x be a point of a set $A \in \mathcal{T}$. Then A has density 1 at x. For each positive integer n, let F_n be an ordinary-closed subset of $(x - 1/2n, x + 1/2n) \cap A$ such that $m(F_n) > (1 - 1/n) m[(x - 1/2n, x + 1/2n) \cap A]$. If $F = \{x\} \cup \bigcup_1^\infty F_n$, then F is an ordinary-closed set, and $\phi(F) \subset F \subset A$. Since A has density 1 at x,

$$nm[(x - 1/2n, x + 1/2n) \cap F] \geq nm(F_n) \to 1 .$$

Therefore F has density 1 at x, and so $x \in \phi(F)$. Thus $\phi(F)$ is a $\mathcal{T}$-neighborhood of x whose $\mathcal{T}$-closure is contained in F, and therefore in A. ☐

Thus Lebesgue measure in any open interval is a category measure relative to the density topology. Lebesgue measure in R is not finite, and therefore not a category measure as we have defined the term.

However, it is easy to define an equivalent finite measure, which is then a category measure relative to the density topology in R. Relative to the density topology, the extended principle of duality is valid, and it is no longer possible to decompose R into a nullset and a set of first category.

The method we have described is not the only way in which category measure spaces can be obtained. The earliest examples to be recognized were the Boolean measure spaces, that is, spaces obtained from finite measure algebras by means of the Stone representation theorem. These provide examples of compact Hausdorff spaces that admit a category measure [13]. Among the continuous images of these spaces may be found still other examples [25]. The study of such spaces belongs more properly to that of Boolean algebras, and so we shall not discuss them here.

Supplementary Notes and Remarks

CHAPTER 3

For an alternative proof of the Lebesgue density theorem (3.20), see [84].

CHAPTER 4

It follows from Theorem 4.8 that if A and B have positive measure, or have the property of Baire and are of second category, then their difference set $D = \{a - b : a \in A, b \in B\}$ contains an interval. [Choose x so that $A \cap (x + B)$ has positive measure or is of second category. Then $(x - \delta, x + \delta) \subset D$ for the corresponding δ.] The two parts of this well-known proposition are due, respectively, to H. Steinhaus and S. Piccard. For generalizations and refinements, see [49] and [72]. When A and B are large in different senses, for instance, when $R - A$ is a nullset and $R - B$ is of first category, the conclusion need not hold [43].

CHAPTER 5

Recent advances in set theory have shed light on the extent to which the axiom of choice (AC) is needed to prove the existence of non-measurable sets. According to Solovay [76], if the existence of inaccessible cardinals is not inconsistent with the axioms of set theory (ZF), there exist models in which all subsets of R are Lebesgue measurable and have the property of Baire, and in which the principle of dependent choices (see below, Chapter 9) is satisfied. This comes close to showing that we cannot hope to define a non-measurable set explicitly, or even by means of an infinite sequence of choices. For a good summary of these and related results, see [45].

On the other hand, the full force of AC is not needed; either the (Boolean) prime ideal theorem or the existence of a choice function for each family of 2-element sets is sufficient [74] [54, Problem 1.10]. Either of these consequences of AC implies that there exists a family $\mathscr{E}$ of subsets of N such that (i) if $A \subset N$, then either $A \in \mathscr{E}$ or $N - A \in \mathscr{E}$ but not both, and (ii) $A \triangle F \in \mathscr{E}$ whenever $A \in \mathscr{E}$ and F is a finite subset of N. The indicator functions of the members of such a family $\mathscr{E}$ constitute a tail set E in the space $X = \mathscr{P} X_i$ defined on page 84. Interchanging 0 and 1 in each X_i defines a measure-preserving homeomorphism T of X onto itself. Since $TE = X - E$, neither E nor $X - E$ can be of measure zero or of first category. It follows from Theorems 21.3 and 21.4 that E cannot be measurable or have the property of Baire. Hence $g(E)$ (page 84) is a subset of $[0, 1]$ that is not Lebesgue measurable and does not have the property of Baire. (Note that the restriction of g to $X - g^{-1}$(dyadic rationals) is a measure-preserving homeomorphism.)

CHAPTER 6

Interesting applications of Mazur's game appear in [51], [55], [56], and [79]. For an abstract version of the game, see [62].

The determinateness for Borel sets of the game studied by Gale and Stewart (page 30), also known as Ulam's game [81, Problem 43], was finally established by D. A. Martin [58]. Although the conclusion appears to be a statement only about Borel subsets of the line, the proof involves sets of very large cardinality. It is remarkable that this feature of the proof is shown to be unavoidable; Borel determinacy is not provable in systems of set theory in which iteration of the power set axiom is restricted.

The possibility of considering the determinateness of certain games as an axiom (inconsistent with AC) has been studied, especially by Mycielski [65].

CHAPTER 8

Egoroff's theorem (8.3) need not hold for a one-parameter family f_t of measurable functions in place of f_n. Walter [82] gives the following counterexample.

Let V be a Vitali set contained in $[0, 1/2)$ and let E be the subset of the strip $S = [0, 1] \times [2, \infty)$ defined by

$$E = \bigcup_{n=2}^{\infty} \{(x, t) \in S : t - x = n, x \in V + 1/n\}.$$

Each horizontal or vertical section of E has at most one point, hence the sections $f_t = \chi_E(\cdot, t)$ of the indicator function of E define a family of measurable functions on $I = [0, 1]$ such that $f_t(x) \to 0$ as $t \to \infty$ for each $x \in I$. If the convergence is uniform on a set $A \subset I$, then $(A \times [p, \infty)) \cap E = \emptyset$ for some integer $p \geqq 2$. Hence $V + 1/p \subset I - A$, therefore $m^*(I - A)$ cannot be smaller than the fixed positive number $m^*(V)$.

CHAPTER 9

The proof of the Baire category theorem (9.1) given on page 41 implicitly uses the axiom of choice in the following weak form, known as the principle of dependent choices (DC).

If $S \neq \emptyset$, $R \subset S \times S$, and for each $x \in S$ there is at least one $y \in S$ such that $(x, y) \in R$, then there is a sequence $\{x_i\} \subset S$ such that $(x_i, x_{i+1}) \in R$ for $i = 1, 2, \ldots$.

It is remarkable that this principle, which is known to be weaker than AC but not provable from ZF alone [54], is actually *logically equivalent* to the Baire category theorem! (see [44]). More precisely, the validity of Baire's theorem in the product of a sequence of copies of an arbitrary discrete space implies DC as follows. Endow S with the discrete topology and let X be the space of mappings $f: N \to S$ with the product topology. Then X is completely metrizable [take $\rho(f, g) = 1/\min\{n: f(n) \neq g(n)\}$]. The hypothesis of DC implies that for each $k \in N$ the open set

$$A_k = \bigcup_{l > k} \bigcup_{(s, t) \in R} \{f \in X : f(k) = s, f(l) = t\}$$

is dense in X. Choose $g \in \bigcap A_k$ and let $k_1 = 1$. If k_i has been defined, let k_{i+1} be the least integer $k_{i+1} > k_i$ such that $(g(k_i), g(k_{i+1})) \in R$ (which exists because $g \in A_{k_i}$). Then k_i is defined for all $i \in N$ and the sequence $x_i = g(k_i)$ satisfies DC.

It should be emphasized, however, that in a complete *separable* metric space the Baire category theorem does not require any form of the axiom of choice (compare the proof of Theorem 1.3).

CHAPTER 11

R. D. Mauldin [59] has recently shown that the set of functions in $C[0, 1]$ that have a finite derivative somewhere in $[0, 1]$ (one-sided at 0 or 1) is actually an analytic, *non-Borel* subset of C. Hence the nowhere differentiable functions constitute a coanalytic non-Borel subset of C.

CHAPTER 13

For a kind of dual of Theorem 13.1, see [80, page 682].

Theorem 13.3(b) says that a function on $[0, 1]$ is topologically equivalent to a Riemann integrable function if and only if it is bounded and continuous except at a set of points of first category. This result may be compared with a theorem of Maximoff [47] which says that a function is topologically equivalent to a derivative if and only if it is a Darboux function of the first class of Baire. (A Darboux function is one that has the intermediate-value property, that is, maps intervals onto intervals.)

CHAPTER 15

As Bruckner [46] observed, Theorem 15.1 implies that if $E \subset R^2$ has the property of Baire, then for each $p \in E$ except a set of first category, and for each line L through p whose inclination does not belong to a certain set of first category in $[0, \pi)$ (depending upon p), the set $L \cap E'$ is of first category at p relative to L. [At any $p \in E$ where E' is of first category, apply 15.1 to $R^2 - \{p\}$ in a polar coordinate representation.] Thus nearly all points of E are points of "linear categorical density" in nearly all directions.

As another application of Theorem 15.1 let $\mathscr{F}$ be a disjoint family of semirays (half lines without endpoint) in R^2 whose union is of first category. If the semirays all have the same direction, or are directed away from some point 0, then 15.1 implies that the set E of endpoints of members of $\mathscr{F}$ is of first category in R^2. However, if the directions of the semirays are unrestricted, it is possible for E to be a residual subset of R^2 [53].

The proof of Theorem 15.4 uses the fact that any plane set of second category having the property of Baire contains a product set $U \times V$ of second category minus a set of first category. The measure-theoretic analogue of this statement is not true: there exists a plane set of positive measure that contains no product set of positive measure minus a nullset [49]. Nevertheless, Theorem 15.4 has a valid measure-theoretic analogue, as is well known [12, §36.A].

CHAPTER 16

The Banach category theorem (16.1) can be formulated in several other ways:

(1) *If $E \subset X$ is of first category at each point of E, then E is of first category.*

(2) *If $E \subset X$ is of second category, then E is of second category at every point of some nonempty open subset of X.*

(3) *If S is the union of a family of sets each open relative to S and of first category in X, then S is of first category in X* [18, §10,III].

It is easy to derive each of these statements from 16.1, and *vice versa*. For example, to deduce (1) or (3), apply 16.1 to the subspace $Y = S \cap S^{-\prime-\prime}$ when $Y \neq \emptyset$ and observe that the category of Y is the same relative to X and to Y. (More generally, if $E \subset Y \subset G \subset X$, then the nowhere denseness and category of E are the same relative to X and to Y in case Y is dense in G and G is an open subset of X.) Version (2) is also a corollary of the Banach–Mazur game theorem (6.1), suitably generalized [9].

As W. Moran observed (personal communication), a simpler example to show that in Theorem 16.4 the hypothesis of metrizability cannot be omitted is the following. Let X be the set of ordinals less than Ω with the order topology and define $\mu(E)$ equal to 1 or 0 according as E or $X - E$ contains an unbounded closed subset of X [12, §52(10)]. The domain of the measure so defined is exactly the algebra of Borel subsets of X [70]. The hypothesis of metrizability *can* be omitted, however, for measures μ that are defined at least for all Borel subsets of a topological space and assign measure zero to every nowhere dense closed set. Such measures, which generalize the category measures of Chapter 22, have been studied by a number of authors and are variously called normal, hyperdiffuse, or residual [42], [50]. [To prove 16.4 for such a measure, let S be the union of a maximal disjoint family $\mathcal{F}$ of open sets of measure zero. Since each subset of $\mathcal{F}$ corresponds to an open set and card $\mathcal{F}$ has measure zero, we must have $\mu(S) = 0$. Hence $\mu(\bar{S}) = 0$ and $\bar{S}$ contains the union of any family of open sets of measure zero. The example on page 64, next to last paragraph, shows that the cardinality restriction in 16.4 is essential even for measures of this special kind.]

CHAPTER 17

A generalized version of Theorem 17.2 holds even when T is multivalued and $I \subset S$ is merely closed under countable union [77].

The assertion (p. 69) that the ergodic theorem does not have a generally valid category-theoretic analogue is meant to be interpreted as follows. Even when T is a measure-preserving homeomorphism of a compact metric space X, there may exist continuous functions for which the orbital (or time) average exists only on a set of first category. By a theorem of Dowker [48], this is the case whenever each orbit is dense in X and T admits more than one normalized invariant Borel measure. [Let f be a continuous function whose integrals relative to two invariant measures are different, and define $F_n(x) = (1/n)\sum_{i=0}^{n-1} f(T^i x)$. By the ergodic theorem and Baire's theorem on functions of first class, there is a point $p \in X$ where the sequence $\{F_n(p)\}$ is divergent. Choose numbers α and β so that $\underline{\lim} F_n(p) < \alpha < \beta < \overline{\lim} F_n(p)$. Then the G_δ set $\{x \in X : \underline{\lim} F_n(x) < \alpha < \beta < \overline{\lim} F_n(x)\}$ contains the orbit of p, is therefore dense, and $\{F_n(x)\}$ diverges everywhere on this set.] For an explicit example of such a system, see [66, §10].

CHAPTER 18

On page 72 it is taken for granted that disjoint plane regions R_k of diameter less than ϵ can always be constructed within the square X so as to separate finitely many pairs of interior points $\{p_k, q_k\}$ provided all of these points are distinct and $|p_k - q_k| < \epsilon$ for each k. For a detailed proof of this geometric fact, see [68]. (In general, disjoint finite subsets of the plane can be so separated if and only if the diameter of each is less than $\epsilon\sqrt{3}/2$ [68].) It follows that if $p_1, \ldots, p_K$ are distinct interior points of X, and likewise $q_1, \ldots, q_K$, and $|p_k - q_k| < \epsilon$ for all k, then $Sp_k = q_k$ $(k = 1, \ldots, K)$ for some S in $M_\epsilon = \{S \in M : \rho(S, \mathrm{id}) < \epsilon\}$. [First displace the points p_k to nearby points r_k distinct from all the points q_k.]

Following S. Alpern [41], an elegant alternative proof that the sets E_{ij} (page 71) are dense in M can be based on the marriage theorem instead of the recurrence theorem. Given $T \in M$ and $\epsilon > 0$, take a dyadic subdivision of X into cells $\sigma_1, \ldots, \sigma_N$ with $\mathrm{diam}\,\sigma_i < \epsilon$ and $\mathrm{diam}\,T\sigma_i < \epsilon$. Since T preserves measure, the images of any k of these cells meet at least k of them. Hence, by the marriage theorem, there exists a permutation P of the σ_i such that $T\sigma_i$ meets $P\sigma_i$ for $i = 1, \ldots, N$. Consider P to act on each σ_i as a translation. Let $p_1, \ldots, p_{4N}$ denote the centers of the cells of the next finer dyadic subdivision of X and assign the four of these points that lie in each σ_i to the four edges of σ_i arbitrarily. As noted above, since $|Tp_j - Pp_j| < 2\epsilon$ we can find an $S_1 \in M_{2\epsilon}$ such that $S_1 Tp_j = Pp_j$ for all j. Label the cycles $C_0, \ldots, C_K$ of P so that for $1 \leq k \leq K$ some cell of C_k has an edge e_k in common with one of the cells of $C_0 \cup \cdots \cup C_{k-1}$. Choose an $S_2 \in M_\epsilon$ that permutes cyclically the points p_j belonging to just one cell of each cycle and leaves the other points p_j fixed, and choose an $S_3 \in M_{2\epsilon}$ that transposes the members of each of the disjoint pairs of points assigned to the edges e_k and leaves the other points p_j fixed. Then $p_1, \ldots, p_{4N}$ constitute a single orbit of $S_3 S_2 S_1 T$. Thus, arbitrarily close to T there is an element of M that belongs to any given one of the sets E_{ij}.

CHAPTER 19

For an application of Theorem 19.5 to some other σ-ideals of sets, see [60].

CHAPTER 20

A Lusin set E not only has Lebesgue measure zero (page 79), it has absolute measure zero; that is, $\mu(E) = 0$ for the completion of each finite nonatomic Borel measure in R. (This follows, for example, from Theorem 16.5.) The existence of sets of absolute measure zero having power $\aleph_1$ can be proved without the continuum hypothesis [75].

Propositions 20.4 and 20.4* can both be viewed as corollaries of a single purely set-theoretic theorem of Sierpiński [73]. For a statement of that theorem and another application of it, see [61].

CHAPTER 21

One way to determine the category of the set E defined on page 85, last paragraph, is to observe that in the Banach-Mazur game, played with blocks of binary digits (page 30), a winning strategy in the game $\langle E, E' \rangle$ is for (B) to choose at each move a block of zeros twice as long as the block of digits just chosen by (A). Then apply Theorem 6.1.

The zero-one law (Theorem 21.4) has been generalized in several directions. See, for example, [57], [63], [69], [71], and [83]. An abstract framework within which a number of analogies between measure and category can be unified is described in [64].

CHAPTER 22

Category measure spaces constructed from lower densities by the method described in this chapter are examples of *topologies in the sense of Hashimoto [52]. (See [67] for other references.) The topology corresponding to a lower density ϕ may be extremally disconnected. (This is the case, for example, when ϕ is a lifting [16, page 59].) However, the Lebesgue density topology makes R connected. [Otherwise there exists a measurable set $A \neq \emptyset, R$, having density 1 at each point of A and density 0 at each point of A'. Choose a number a that is not an upper bound of A or A' and for $a < x < \infty$ define $f(x) = m([a, x] \cap A)$. Then $f'(x) = \chi_A(x)$ on (a, ∞) and the range of f' is $\{0, 1\}$, contrary to the intermediate-value theorem for derivatives.] For additional properties of the density topology, see [78].

References

1. Banach, S.: Sur les suites d'ensembles excluant l'existence d'une mesure. Colloq. Math. **1**, 103—108 (1948).
2. Besicovitch, A. S.: A problem on topological transformation of the plane. Fund. Math. **28**, 61—65 (1937).
3. Birkhoff, G. D.: Dynamical systems. Amer. Math. Soc. Colloq. Publ., Vol. 9, New York 1927.
4. Borel, E.: Leçons sur les fonctions de variables réeles. Paris: Gauthier-Villars 1905.
5. — Leçons sur la theorie des fonctions. Paris: Gauthier-Villars 1914.
6. Brouwer, L. E. J.: Lebesguesches Mass und Analysis Situs. Math. Ann. **79**, 212—222 (1919).
7. Carathéodory, C.: Über den Wiederkehrsatz von Poincaré. S. B. Preuss. Akad. Wiss. **1919**, 580—584.
8. Contributions to the theory of games, Vol. 2. Annals of Math. Studies, no. 28, pp. 245—266. Princeton, N. J.: Princeton Univ. Press 1953.
9. Contributions to the theory of games, Vol. 3. Annals of Math. Studies, no. 39, pp. 159—163. Princeton, N. J.: Princeton Univ. Press 1957.
10. Erdös, P.: Some remarks on set theory. Ann. of Math. (2) **44**, 643—646 (1943).
11. Goffman, C., Neugebauer, C. J., Nishiura, T.: Density topology and approximate continuity. Duke Math. J. **28**, 497—505 (1961).
12. Halmos, P. R.: Measure theory. New York: D. Van Nostrand 1950.
13. — Lectures on Boolean algebras. Van Nostrand Math. Studies, no. 1. Princeton, N. J.: D. Van Nostrand 1963.
14. Hilmy, H.: Sur les théorèmes de récurrence dans la dynamique générale. Amer. J. Math. **61**, 149—160 (1939).
15. Ionescu Tulcea, A., Ionescu Tulcea, C.: On the lifting property I. J. Math. Anal. Appl. **3**, 537—546 (1961).
16. — — Topics in the theory of lifting. Ergebnisse der Math. und ihrer Grenzgebiete, Bd. 48. Berlin-Heidelberg-New York: Springer 1969.
17. Kemperman, J. H. B., Maharam, D.: R^c is not almost Lindelöf. Proc. Amer. Math. Soc. **24**, 772—773 (1970).
18. Kuratowski, C.: Topologie, Vol. 1, 4th ed. Monografie Matematyczne, Vol. 20. Warszawa: Panstwowe Wydawnictwo Naukowe, 1958.
19. Maharam, D.: On a theorem of von Neumann. Proc. Amer. Math. Soc. **9**, 987—994 (1958).
20. Marczewski, E., Sikorski, R.: Measures in non-separable metric spaces. Colloq. Math. **1**, 133—139 (1948).
21. — — Remarks on measure and category. Colloq. Math. **2**, 13—19 (1949).
22. Mycielski, J.: Some new ideals of sets on the real line. Colloq. Math. **20**, 71—76 (1969).

23. Oxtoby, J. C.: The category and Borel class of certain subsets of $\mathscr{L}_p$. Bull. Amer. Math. Soc. **43**, 245—248 (1937).
24. — Note on transitive transformations. Proc. Nat. Acad. Sci. U.S. **23**, 443—446 (1937).
25. — Spaces that admit a category measure. J. Reine Angew. Math. **205**, 156—170 (1961).
26. — Cartesian products of Baire spaces. Fund. Math. **49**, 157—166 (1961).
27. — Ulam, S. M.: On the equivalence of any set of first category to a set of measure zero. Fund. Math. **31**, 201—206 (1938).
28. — — Measure-preserving homeomorphisms and metrical transitivity. Ann. of Math. (2) **42**, 874—920 (1941).
29. Poincaré, H.: Les méthodes nouvelles de la mécanique céleste, Vol. 3. Paris: Gauthier-Villars 1899.
30. Rademacher, H.: Eineindeutige Abbildungen und Meßbarkeit. Monatsh. Math. Phys. **27**, 183—291 (1916).
31. Riesz, F., Sz.-Nagy, B.: Leçons d'analyse fonctionnelle, 4th ed. Paris: Gauthier-Villars 1965.
32. Shoenfield, J. R.: Mathematical logic. Reading, Mass.: Addison-Wesley 1967.
33. Sierpinski, W.: Sur une extension de la notion de l'homéomorphie. Fund. Math. **22**, 270—275 (1934).
34. — Sur la dualité entre la première catégorie et la mesure nulle. Fund. Math. **22**, 276—280 (1934).
35. — Hypothèse du continu. Monografie Matematyczne, Vol. 4. Warszawa-Lwów 1934.
36. — Lusin, N.: Sur une décomposition d'un intervalle en une infinité non dénombrable d'ensembles non mesurables. C. R. Acad. Sci. Paris **165**, 422—424 (1917).
37. Sikorski, R.: On an unsolved problem from the theory of Boolean algebras. Colloq. Math. **2**, 27—29 (1949).
38. Szpilrajn, E.: Remarques sur les fonctions complètement additives d'ensemble et sur les ensembles joissant de la propriété de Baire. Fund. Math. **22**, 303—311 (1934).
39. Ulam, S. M.: Zur Maßtheorie in der allgemeinen Mengenlehre. Fund. Math. **16**, 141—150 (1930).
40. — A collection of mathematical problems. New York: Interscience 1960.

Supplementary References

41. Alpern, S.: A new proof of the Oxtoby-Ulam theorem. Ergodic Theory; a Seminar (J. Moser, ed.). Courant Institute, New York University 1975, pp. 125—131.
42. Armstrong, T. E., Prikry, K.: Residual measures. Illinois J. Math. **22**, 64—78 (1978).
43. Bagemihl, F.: Some sets of sums and differences. Michigan Math. J. **4**, 289—290 (1957).
44. Blair, C. E.: The Baire category theorem implies the principle of dependent choices. Bull. Acad. Polon. Sci. **25**, 933—934 (1977).
45. Briggs, J. M., Schaffter, T.: Measure and cardinality. Amer. Math. Monthly **86**, 852—855 (1979).

46. Bruckner, A. M.: A category analogue of a theorem on metric density. Israel J. Math. **11**, 249—253 (1972).
47. — Leonard, J. L.: Derivatives. Amer. Math. Monthly **73**, H. E. Slaught Memorial Papers, No. 11, 24—56 (1966).
48. Dowker, Y. N.: The mean and transitive points of homeomorphisms. Ann. of Math. (2) **58**, 123—133 (1953).
49. Erdös, P., Oxtoby, J. C.: Partitions of the plane into sets having positive measure in every non-null measurable product set. Trans. Amer. Math. Soc. **79**, 91—102 (1955).
50. Flachsmeyer, J., Lotz, S.: A survey on hyperdiffuse measures. I. Proceedings of Conference Topology and Measure (Zinnowitz, 1974), Part I, pp. 87—128. Ernst-Moritz-Arndt Univ., Greifswald, 1978.
51. Fleissner, W. G., Kunen, K.: Barely Baire spaces. Fund. Math. **101**, 229—240 (1978).
52. Hashimoto, H.: On the *topology and its application. Fund. Math. **101**, 5—10 (1976).
53. Humke, P. D.: Baire category and disjoint rectilinear accessibility. J. London Math. Soc. (2) **14**, 245—248 (1976).
54. Jech, T. J.: The axiom of choice. Amsterdam-London: North-Holland 1973.
55. Krom, M. R.: Cartesian products of metric Baire spaces. Proc. Amer. Math. Soc. **42**, 588—594 (1974).
56. — Infinite games and special Baire space extensions. Pacific J. Math. **55**, 483—487 (1974).
57. Kuratowski, K.: On the concept of strongly transitive systems in topology. Ann. Mat. Pura Appl. (4) **98**, 357—363 (1974).
58. Martin, D. A.: Borel determinacy. Ann of Math. (2) **102**, 363—371 (1975).
59. Mauldin, R. D.: The set of continuous nowhere differentiable functions. Pacific J. Math. **83**, 199—205 (1979).
60. Mendez, C. G.: On the Sierpiński-Erdös and the Oxtoby-Ulam theorems for some new sigma-ideals of sets. Proc. Amer. Math. Soc. **72**, 182—188 (1978).
61. Miller, H. I.: Some results connected with a problem of Erdös. II. Proc. Amer. Math. Soc. **75**, 265—268 (1979).
62. Morgan, J. C., II: Infinite games and singular sets. Colloq. Math. **29**, 7—17 (1974).
63. — On zero-one laws. Proc. Amer. Math. Soc. **62**, 353—358 (1977).
64. — Baire category from an abstract viewpoint. Fund. Math. **94**, 13—23 (1977).
65. Mycielski, J.: On the axiom of determinateness. I, II. Fund. Math. **53**, 205—224 (1964); **59**, 203—212 (1966).
66. Oxtoby, J. C.: Ergodic sets. Bull. Amer. Math. Soc. **58**, 116—136 (1952).
67. — The kernel operation on subsets of a T_1-space. Fund. Math. **90**, 275—284 (1976).
68. — Diameters of arcs and the gerrymandering problem. Amer. Math. Monthly **84**, 155—162 (1977).
69. Rao, K. P. S. Bhaskara, Pol, R.: Topological zero-one laws. Colloq. Math. **39**, 13—23 (1978).
70. Rao, M. Bhaskara, Rao, K. P. S. Bhaskara: Borel σ-algebra on $[0, \Omega]$. Manuscripta Math. **5**, 195—198 (1971).
71. — — A category analogue of the Hewitt-Savage zero-one law. Proc. Amer. Math. Soc. **44**, 497—499 (1974).

72. Sander, W.: A generalization of a theorem of S. Piccard. Proc. Amer. Math. Soc. **73**, 281—282 (1979).
73. Sierpiński, W.: Un théorème de la théorie générale des ensembles et ses applications. C. R. Soc. Sci. Varsovie **28**, 131—135 (1936).
74. — Fonctions additives non complètement additives et fonctions non mesurables. Fund. Math. **30**, 96—99 (1938).
75. — Szpilrajn (= Marczewski), E.: Remarque sur le problème de la mesure. Fund. Math. **26**, 256—261 (1936).
76. Solovay, R. M.: A model of set theory in which every set of reals is Lebesgue measurable. Ann of Math. (2) **92**, 1—56 (1970).
77. Sucheston, L.: A note on conservative transformations and the recurrence theorem. Amer. J. Math. **79**, 444—447 (1957).
78. Tall, F. D.: The density topology. Pacific J. Math. **62**, 275—284 (1976).
79. Telgársky, R.: On some topological games. Proc. Fourth Prague Topological Sympos., Prague 1976, Part B, 461—472.
80. Ulam, S. M.: Sets, numbers, and universes: selected works. Cambridge, Mass.: MIT Press 1974.
81. — The Scottish book: A LASL Monograph. Los Alamos Scientific Lab. 1977. LA-6832.
82. Walter, W.: A counterexample in connection with Egorov's theorem. Amer. Math. Monthly **84**, 118—119 (1977).
83. White, H. E. Jr.: Two unrelated results involving Baire spaces. Proc. Amer. Math. Soc. **44**, 463—466 (1974).
84. Zajíček, L.: An elementary proof of the one-dimensional density theorem. Amer. Math. Monthly **86**, 297—298 (1979).

Index

Graduate Texts in Mathematics

Soft and hard cover editions are available for each volume up to vol. 14, hard cover only from Vol. 15

CPSIA information can be obtained
at www.ICGtesting.com
Printed in the USA
FFOW03n1354180316
22462FF